Dowb N F. Power

The Culture of
Technology

The Culture of Technology

Arnold Pacey

The MIT Press
Cambridge, Massachusetts

First MIT Press edition, 1983

Library of Congress Cataloging in Publication Data
Pacey, Arnold.
    The culture of technology.

    Bibliography: p.
    Includes index.
    1. Technology. I. Title.
T49.5.P3   1983   306'.46   83-11393
ISBN 0-262-16093-5

Printed and bound in Great Britain.

Contents

Acknowledgements   vii

1 Technology: Practice and Culture   1

2 Beliefs about Progress   13

3 The Culture of Expertise   35

4 Beliefs about Resources   55

5 Imperatives and Creative Culture   78

6 Women and Wider Values   97

7 Value-conflicts and Institutions   120

8 Innovative Dialogue   137

9 Cultural Revolution   160

Notes   180

Select Bibliography   200

Index   203

Acknowledgements

This book offers a somewhat personal view of technology for which only I can be blamed; but equally, it offers a wide-ranging view, and that has put me in great debt to colleagues with broader experience than mine. There has also been a collective stimulus from many people in the organizations for which I have worked during the last decade, especially the Open University and OXFAM, but also the Institute of Development Studies (University of Sussex), the Intermediate Technology Development Group, and the London School of Hygiene and Tropical Medicine.

Individual debts are always difficult to identify with justice, but I am conscious that David Bradley and Charles Webster pointed me toward much of the most valuable of this experience. Another particularly valuable form of help has come from those who, through prompting, loans and gifts, have caused me to fill some of the gaps in my always patchy reading, including Tony Barker, Alice Crampin, Jacques Grinevald (Geneva), Roger Newton, Philip Pacey, Simon Watt, and Penny Williams (Calgary). Those who have helped by reading chapters in draft, and often writing lengthy comments, are Angus Buchanan, Shione Carden, Howard Erskine Hill, David Farrar, Sylvia Farrar, John Naughton, Alison Ravetz, Jerry Ravetz, Penni Thompson and Charles Webster.

For permission to reproduce diagrams and tables, I am indebted to Intermediate Technology Publications Ltd (figures 2 and 7), Unilever PLC and *Progress* magazine (figure 3a), G. T. Shepherd (figure 8), Linda Richardson (figure 9) and the Sarvodaya Shramadana Movement, Moratuwa, Sri Lanka (table 6).

The many quotations in the text are fully acknowledged in the notes at the end of the book. With regard to the longer passages used, I am grateful to the following:

The Editor, *Food Policy*, Butterworth Scientific Ltd, PO Box 63, Westbury House, Bury Street, Guildford, Surrey, for permission to quote from Davidson R. Gwatkin, 'Food policy, nutrition planning and survival – the cases of Kerala and Sri Lanka', in vol. 4, no. 4, November 1979, pp. 245–58.

Langley Technical Services, 95 Sussex Place, Slough, for permission to quote from *Architect or Bee?* by Mike Cooley, copyright 1980.

Ministry of Supply and Services, Ottawa, Canada, for permission to quote from *Northern Frontier, Northern Homeland: the Report of the Mackenzie Valley Pipeline Inquiry*, by Mr. Justice Thomas R. Berger, copyright 1977, Ministry of Supply and Services.

OXFAM, 274 Banbury Road, Oxford, for permission to quote from *Against the Grain*, by Tony Jackson, copyright 1982 OXFAM.

Philip Pacey, for permission to quote from *In the Elements Free* (Newcastle: Galloping Dog Press), copyright 1983 by Philip Pacey.

Routledge and Kegan Paul Ltd., London, and Stanford University Press, Stanford, California, for permission to quote from *The New Liberty*, copyright 1975 by Ralf Dahrendorf.

St. Martin's Press, Inc., New York, for permission to quote from *The Existential Pleasures of Engineering*, by Samuel C. Florman, copyright 1976 Samuel C. Florman.

SCM Press Ltd, London, for permission to quote from *Transport of Delight*, copyright 1976 by Jack Burton.

John Wiley & Sons, Chichester, for permission to quote from *Catastrophe and Cornucopia*, copyright 1982 by Stephen Cotgrove.

Verso Edition and New Left Books, London, for permission to quote from *Problems in Materialism and Culture*, copyright 1980 by Raymond Williams.

Arnold Pacey
Oxford
31 December 1982

1
Technology:
Practice and Culture

Questions of neutrality

Winter sports in North America gained a new dimension during the 1960s with the introduction of the snowmobile. Ridden like a motorcycle, and having handlebars for steering, this little machine on skis gave people in Canada and the northern United States extra mobility during their long winters. Snowmobile sales doubled annually for a while, and in the boom year of 1970–1 almost half a million were sold. Subsequently the market dropped back, but snowmobiling had established itself, and organized trails branched out from many newly prosperous winter holiday resorts. By 1978, there were several thousand miles of public trails, marked and maintained for snowmobiling, about half in the province of Quebec.

Although other firms had produced small motorized toboggans, the type of snowmobile which achieved this enormous popularity was only really born in 1959, chiefly on the initiative of Joseph-Armand Bombardier of Valcourt, Quebec.[1] He had experimented with vehicles for travel over snow since the 1920s, and had patented a rubber-and-steel crawler track to drive them. His first commercial success, which enabled his motor repair business to grow into a substantial manufacturing firm, was a machine capable of carrying seven passengers which was on the market from 1936. He had other successes later, but nothing that caught the popular imagination like the little snowmobile of 1959, which other manufacturers were quick to follow up.

However, the use of snowmobiles was not confined to the North American tourist centres. In Sweden, Greenland and the Canadian Arctic, snowmobiles have now become part of the equipment on which many communities depend for their livelihood. In Swedish Lapland they are used for reindeer herding. On Canada's Banks Island they have enabled Eskimo trappers to continue providing their families' cash income from the traditional winter harvest of fox furs.

Such use of the snowmobile by people with markedly different cultures may seem to illustrate an argument very widely advanced in discussions of problems associated with technology. This is the argument which states that technology is culturally, morally and politically neutral – that it provides tools independent of local value-systems which can be used impartially to support quite different kinds of lifestyle.

Thus in the world at large, it is argued that technology is 'essentially amoral, a thing apart from values, an instrument which can be used for good or ill'.[2] So if people in distant countries starve; if infant mortality within the inner cities is persistently high; if we feel threatened by nuclear destruction or more insidiously by the effects of chemical pollution, then all that, it is said, should not be blamed on technology, but on its misuse by politicians, the military, big business and others.

The snowmobile seems the perfect illustration of this argument. Whether used for reindeer herding or for recreation, for ecologically destructive sport, or to earn a basic living, it is the same machine. The engineering principles involved in its operation are universally valid, whether its users are Lapps or Eskimos, Dene (Indian) hunters, Wisconsin sportsmen, Quebecois vacationists, or prospectors from multinational oil companies. And whereas the snowmobile has certainly had a social impact, altering the organization of work in Lapp communities, for example, it has not necessarily influenced basic cultural values. The technology of the snowmobile may thus appear to be something quite independent of the lifestyles of Lapps or Eskimos or Americans.

One look at a modern snowmobile with its fake streamlining and flashy colours suggests another point of view. So does the advertising which portrays virile young men riding the machines with sexy companions, usually blonde and usually riding pillion. The Eskimo who takes a snowmobile on a long expedition in the Arctic quickly discovers more significant discrepancies. With his traditional means of transport, the dog-team and sledge, he could refuel as he went along by hunting for his dogs' food. With the snowmobile he must take an ample supply of fuel and spare parts; he must be skilled at doing his own repairs and even then he may take a few dogs with him for emergency use if the machine breaks down. A vehicle designed for leisure trips between well-equipped tourist centres presents a completely different set of servicing problems when used for heavier work in more remote areas. One Eskimo 'kept his machine in his tent so it could be warmed up before starting in the morning, and even then was plagued by mechanical failures'.[3] There are stories of other Eskimos, whose mechanical aptitude is well known, modifying their machines to adapt them better to local use.

So is technology culturally neutral? If we look at the construction of a basic machine and its working principles, the answer seems to be yes. But if we look at the web of human activities surrounding the machine, which include its practical uses, its role as a status symbol, the supply of fuel and spare parts, the organized tourist trails, and the skills of its owners, the answer is clearly no. Looked at in this second way, technology is seen as a part of life, not something that can be kept in a separate compartment. If it is to be of any use, the snowmobile must fit into a pattern of activity which belongs to a particular lifestyle and set of values.

The problem here, as in much public discussion, is that 'technology' has become a catchword with a confusion of different meanings. Correct usage of the word in its original sense seems almost beyond recovery, but consistent distinction between different levels of meaning is both possible and necessary. In medicine, a distinction of the kind required is often made by talking about 'medical practice' when a general term is required, and employing the phrase 'medical science' for the more strictly technical aspects of the subject. Sometimes, references to 'medical practice' only denote the organization necessary to use medical knowledge and skill for treating patients. Sometimes, however, and more usefully, the term refers to the whole activity of medicine, including its basis in technical knowledge, its organization, and its cultural aspects. The latter comprise the doctor's sense of vocation, his personal values and satisfactions, and the ethical code of his profession. Thus 'practice' may be a broad and inclusive concept.

Once this distinction is established, it is clear that although medical practice differs quite markedly from one country to another, medical science consists of knowledge and techniques which are likely to be useful in many countries. It is true that medical science in many western countries is biased by the way that most research is centred on large hospitals. Even so, most of the basic knowledge is widely applicable and relatively independent of local cultures. Similarly, the design of snowmobiles reflects the way technology is practised in an industrialized country – standardized machines are produced which neglect some of the special needs of Eskimos and Lapps. But one can still point to a substratum of knowledge, technique and underlying principle in engineering which has universal validity, and which may be applied anywhere in the world.

We would understand much of this more clearly, I suggest, if the concept of practice were to be used in all branches of technology as it has traditionally been used in medicine. We might then be better able to see which aspects of technology are tied up with cultural values, and which aspects are, in some respects, value-free. We would be better able to appreciate technology as a human activity and as part of life. We might then see it not only as comprising machines, techniques and crisply precise knowledge, but also as involving characteristic patterns of organization and imprecise values.

Medical practice may seem a strange exemplar for the other technologies, distorted as it so often seems to be by the lofty status of the doctor as an expert. But what is striking to anybody more used to engineering is that medicine has at least got concepts and vocabulary which allow vigorous discussion to take place about different ways of serving the community. For example, there are phrases such as 'primary health care' and 'community medicine' which are sometimes emphasized as the kind of medical practice to be encouraged wherever the emphasis on hospital medicine has been pushed too far. There are also some interesting adaptations of the language of medical practice. In parts of Asia, para-medical workers, or para-medics, are now paralleled by 'para-agros' in agriculture, and the Chinese barefoot doctors have inspired the suggestion that barefoot technicians could be recruited to deal with urgent problems in village water supply. But despite these occasional borrowings, discussion about practice in most branches of technology has not progressed very far.

Problems of definition

In defining the concept of technology-practice more precisely, it is necessary to think with some care about its human and social aspect. Those who write about the social relations and social control of technology tend to focus particularly on organization. In particular, their emphasis is on planning and administration, the management of research, systems for regulation of pollution and other abuses, and professional organization among scientists and technologists. These are important topics, but there is a wide range of other human content in technology-practice which such studies often neglect, including personal values and individual experience of technical work.

To bring all these things into a study of technology-practice may seem likely to make it bewilderingly comprehensive. However, by remembering the way in which medical practice has a technical and ethical as well as an organizational element, we can obtain a more orderly view of what technology-practice entails. To many politically-minded people, the *organizational aspect* may seem most crucial. It represents many facets of administration, and public policy; it relates to the activities of designers, engineers, technicians, and production workers, and also concerns the users and consumers of whatever is produced. Many other people, however, identify technology with its *technical aspect*, because that has to do with machines, techniques, knowledge and the essential activity of making things work.

Beyond that, though, there are values which influence the creativity of designers and inventors. These, together with the various beliefs and habits of thinking which are characteristic of technical and scientific activity, can be indicated by talking about an ideological or *cultural aspect* of technology-practice. There is some risk of ambiguity here, because strictly speaking, ideology, organization, technique and tools are all aspects of the culture of a society. But in common speech, culture refers to values, ideas and creative activity, and it is convenient to use the term with this meaning. It is in this sense that the title of this book refers to the cultural aspect of technology-practice.

All these ideas are summarized by Figure 1, in which the whole triangle stands for the concept of technology-practice and the corners represent its organizational, technical and cultural aspects. This diagram is also intended to illustrate how the word technology is sometimes used by people in a restricted sense, and sometimes with a more general meaning. When technology is discussed in the more restricted way, cultural values and organizational factors are regarded as external to it. Technology is then identified entirely with its technical aspects, and the words 'technics' or simply 'technique' might often be more appropriately used. The more general meaning of the word, however, can be equated with technology-practice, which clearly is not value-free and politically neutral, as some people say it should be.

Some formal definitions of technology hover uncertainly between the very general and the more restricted usage. Thus J. K. Galbraith defines technology as 'the systematic application of scientific or other organized knowledge to practical tasks'.[4] This sounds a fairly narrow definition, but on reading further one finds that Galbraith thinks of technology as an activity involving complex organizations and value-systems. In view of this, other authors have extended Galbraith's wording.

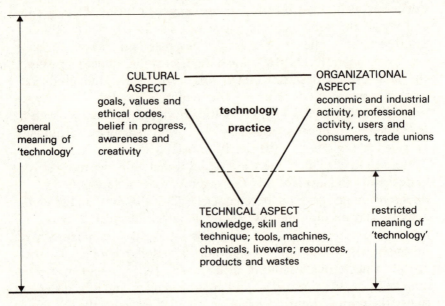

FIGURE 1   *Diagrammatic definitions of 'technology' and 'technology practice'*

For them a definition which makes explicit the role of people and organizations as well as hardware is one which describes technology as 'the application of scientific and other organized knowledge to practical tasks by . . . ordered systems that involve people and machines'.[5] In most respects, this sums up technology-practice very well. But some branches of technology deal with processes dependent on living organisms. Brewing, sewage treatment and the new biotechnologies are examples. Many people also include aspects of agriculture, nutrition and medicine in their concept of technology. Thus our definition needs to be enlarged further to include 'liveware' as well as hardware; technology-practice is thus *the application of scientific and other knowledge to practical tasks by ordered systems that involve people and organizations, living things and machines.*

This is a definition which to some extent includes science within technology. That is not, of course, the same as saying that science is merely one facet of technology with no purpose of its own. The physicist working on magnetic materials or semiconductors may have an entirely abstract interest in the structure of matter, or in the behaviour of electrons in solids. In that sense, he may think of himself as a pure scientist, with no concern at all for industry and technology. But it is no coincidence that the magnetic materials he works on are precisely those that are used in transformer cores and computer memory devices, and that the semiconductors investigated may be used in microprocessors. The scientist's choice of research subject is inevitably influenced by technological requirements, both through material pressures and also via a climate of opinion about what subjects are worth pursuing. And a great deal of science is like this, with goals that are definitely outside technology-practice, but with a practical function within it.

Given the confusion that surrounds usage of the word 'technology', it is not surprising that there is also confusion about the two adjectives 'technical' and 'technological'. Economists make their own distinction, defining change of technique as a development based on choice from a range of known methods, and technological change as involving fundamentally new discovery or invention. This can lead to a distinctive use of the word 'technical'. However, I shall employ this adjective when I am referring solely to the technical aspects of practice as defined by figure 1. For example, the application of a chemical water treatment to counteract river pollution is described here as a 'technical fix' (not a 'technological fix'). It represents an attempt to solve a problem by means of technique alone, and ignores possible changes in practice that might prevent the dumping of pollutants in the river in the first place.

By contrast, when I discuss developments in the practice of technology which include its organizational aspects, I shall describe these as 'technological developments', indicating that they are not restricted to technical form. The terminology that results from this is usually consistent with everyday usage, though not always with the language of economics.

Exposing background values

One problem arising from habitual use of the word technology in its more restricted sense is that some of the wider aspects of technology-practice have come to be entirely forgotten. Thus behind the public debates about resources and the environment, or about world food supplies, there is a tangle of unexamined beliefs and values, and a basic confusion about what technology is for. Even on a practical level, some projects fail to get more than half way to solving the problems they address, and end up as unsatisfactory technical fixes, because important organizational factors have been ignored. Very often the users of equipment (figure 2) and their patterns of organization are largely forgotten.

Part of the aim of this book is to strip away some of the attitudes that restrict our view of technology in order to expose these neglected cultural aspects. With the snowmobile, a first step was to look at different ways in which the use and maintenance of the machine is organized in different communities. This made it clear that a machine designed in response to the values of one culture needed a good deal of effort to make it suit the purposes of another.

A further example concerns the apparently simple hand-pumps used at village wells in India. During a period of drought in the 1960s, large power-driven drilling rigs were brought in to reach water at considerable depths in the ground by means of bore-holes. It was at these new wells that most of the hand-pumps were installed. By 1975 there were some 150,000 of them, but surveys showed that at any one time as many as two-thirds had broken down. New pumps sometimes failed within three or four weeks of installation. Engineers identified several faults, both in the design of the pumps and in standards of manufacture. But although these defects were corrected, pumps continued to go wrong. Eventually it was realized that the breakdowns were not solely an engineering problem. They were also partly an administrative or management issue, in that arrangements for servicing the pumps were not very effective. There was another difficulty, too, because in many villages, nobody felt any personal responsibility for looking after the pumps. It was only when these things were tackled together that pump performance began to improve.

This episode and the way it was handled illustrates very well the importance of an integrated appreciation of technology-practice. A breakthrough only came when all aspects of the administration, maintenance and technical design of the pump were thought out in relation to one another. What at first held up solution of the problem was a view of technology which began and ended with the machine – a

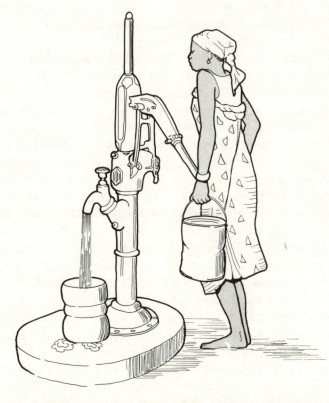

FIGURE 2   *Technology is about 'systems that involve people and machines', and many of the people concerned are users of machines such as hand-pumps or snowmobiles*

view which, in another similar context, has been referred to as tunnel vision in engineering.

Any professional in such a situation is likely to experience his own form of tunnel vision. If a management consultant had been asked about the hand-pumps, he would have seen the administrative failings of the maintenance system very quickly, but might not have recognized that mechanical improvements to the pumps were required. Specialist training inevitably restricts people's approach to problems. But tunnel vision in attitudes to technology extends far beyond those who have had specialized training; it also affects policy-making, and influences popular expectations. People in many walks of life tend to focus on the tangible, technical aspect of any practical problem, and then to think that the extraordinary capabilities of modern technology ought to lead to an appropriate 'fix'. This attitude seems to apply to almost everything from inner city decay to military security, and from pollution to a cure for cancer. But all these issues have a social component. To hope for a technical fix for any of them that does not also involve social and cultural measures is to pursue an illusion.

So it was with the hand-pumps. The technical aspect of the problem was exemplified by poor design and manufacture. There was the organizational difficulty about maintenance. Also important, though, was the cultural aspect of technology as it was practised by the engineers involved. This refers, firstly, to the engineers' way of thinking, and the tunnel vision it led to; secondly, it indicates conflicts of values between highly trained engineers and the relatively uneducated people of the Indian countryside whom the pumps were meant to benefit. The local people probably had exaggerated expectations of the pumps as the products of an all-powerful, alien technology, and did not see them as vulnerable bits of equipment needing care in use and protection from damage; in addition, the local people would have their own views about hygiene and water use.

Many professionals in technology are well aware that the problems they deal with have social implications, but feel uncertainty about how these should be handled. To deal only with the technical detail and leave other aspects on one side is the easier option, and after all, is what they are trained for. With the hand-pump problem, an important step forward came when one of the staff of a local water development unit started looking at the case-histories of individual pump breakdowns. It was then relatively easy for him to pass from a technical review of components which were worn or broken to looking at the social context of each pump. He was struck by the way some pumps had deteriorated but others had not. One well-cared-for pump was locked up during certain hours; another was used by the family of a local official; others in good condition were in places where villagers had mechanical skills and were persistent with improvised repairs. It was these specific details that enabled suggestions to be made about the reorganization of pump maintenance.[6]

A first thought prompted by this is that a training in science and technology tends to focus on general principles, and does not prepare one to look for specifics in quite this way. But the human aspect of technology – its organization and culture – is not easily reduced to general principles, and the investigator with an eye for significant detail may sometimes learn more than the professional with a highly systematic approach.

A second point concerns the way in which the cultural aspect of technology-practice tends to be hidden beneath more obvious and more practical issues. Behind the tangible aspect of the broken hand-pumps lies an administrative problem concerned with maintenance. Behind that lies a problem of political will – the official whose family depended on one of the pumps was somehow well served. Behind that again were a variety of questions concerning cultural values regarding hygiene, attitudes to technology, and the outlook of the professionals involved.

This need to strip away the more obvious features of technology-practice to expose the background values is just as evident with new technology in western countries. Very often concern will be expressed about the health risk of a new device when people are worried about more intangible issues, because health risk is partly a technical question that is easy to discuss openly. A relatively minor technical problem affecting health may thus become a proxy for deeper worries about the way technology is practised which are more difficult to discuss.

An instance of this is the alleged health risks associated with visual display units (VDUs) in computer installations. Careful research has failed to find any real hazard except that operators may suffer eyestrain and fatigue. Yet complaints about more serious problems continue, apparently because they can be discussed seriously with employers while misgivings about the overall systems are more difficult to raise. Thus a negative reaction to new equipment may be expressed in terms of a fear of 'blindness, sterility, etc.', because in our society, this is regarded as a legitimate reason for rejecting it. But to take such fears at face value will often be to ignore deeper, unspoken anxieties about 'deskilling, inability to handle new procedures, loss of control over work'.[7]

Here, then, is another instance where, beneath the overt technical difficulty there are questions about the organizational aspect of technology – especially the organization of specific tasks. These have political connotations, in that an issue about control over work raises questions about where power lies in the work-place, and perhaps ultimately, where it lies within industrial society. But beyond arguments of that sort, there are even more basic values about creativity in work and the relationship of technology and human need.

In much the same way as concern about health sometimes disguises work-place issues, so the more widely publicized environmental problems may also hide underlying organizational and political questions. C. S. Lewis once remarked that 'Man's power over Nature often turns out to be a power exerted by some men over other men with Nature as its instrument', and a commentator notes that this, 'and not the environmental dilemma as it is usually conceived', is the central issue for technology.[8] As such, it is an issue whose political and social ramifications have been ably analysed by a wide range of authors.[9]

Even this essentially political level of argument can be stripped away to reveal another cultural aspect of technology. If we look at the case made out in favour of almost any major project – a nuclear energy plant, for example – there are nearly always issues concerning political power behind the explicit arguments about tangible benefits and costs. In a nuclear project, these may relate to the power of management over trade unions in electricity utilities; or to prestige of governments and the power of their technical advisers. Yet those who operate these levers of power are able to do so partly because they can exploit deeper values relating to the so-called technological imperative, and to the basic creativity that makes innovation possible. This, I argue, is a central part of the culture of technology, and its analysis occupies several chapters in this book. If these values underlying the technological imperative are understood, we may be able to see that here is a stream of feeling which politicians can certainly manipulate at times, but which is stronger than their short-term purposes, and often runs away beyond their control.

2
Beliefs about Progress

Measuring progress

It is understandable that in thinking about particular machines – hand-pumps or VDUs – we habitually focus on hardware rather than human activity. It makes less sense, though, to think like this about more general concepts such as 'progress'. Yet there is a long history of identifying the overall progress of technology with specific inventions or with other strictly technical advances. As early as 1600, inventions such as the printing press, the magnetic compass and firearms were being quoted as evidence of technical progress; more recently, the steam engine and electric light have been added to the list.

There has also been a growing interest in measurable factors which allow key aspects of progress to be expressed numerically and displayed by means of graphs. Some authors have used this method to show how scientific knowledge is cumulative in time,[1] or have plotted the improving performance of specific types of machine. Others have tried to form a composite picture of how technology develops by superimposing data that relate to a number of different techniques. Thus Wedgwood Benn presents a schematic graph showing how computing, transport, communications and weapons have developed, describing this as 'a citizen's guide to the history of technology'.[2] Chauncey Starr goes further and quotes a technological index which combines factors relating to energy efficiency, steel output, communications and skilled manpower in science and engineering.[3]

One striking feature of these studies is that many types of machine appear to show steady development over very long periods of time. The improving accuracy of clocks is a favourite example with many commentators, because a consistent improvement in timekeeping can be plotted over nearly four centuries. Nicholas Rescher, a keen exponent of statistics and diagrams as representations of progress, comments that this kind of consistency in technical improvement is a common finding. 'Again and again – in one science-correlative branch of technology after another', a smooth curve is observed when 'performance-improvement' is plotted as a graph.[4]

These ways of thinking about progress have very serious weaknesses, however. They tend to be over-selective, and lead us to overlook the fact that improvements in one dimension are sometimes accompanied by less desirable developments elsewhere. In agriculture, for example, the amount of food produced can be judged in relation to land, labour or energy. In Britain, as in other western countries, grain output has increased enormously in the present century, especially in relation to the area of land cultivated (figure 3) and the number of people employed. But grain output per unit of energy consumed on farms has *decreased*.[5]

How progress is evaluated depends on circumstances and, in particular, on whether there is a shortage of land or of energy. In small, heavily populated countries such as Britain, and more especially, the Netherlands, land is a valuable commodity, and priority is given to pushing up grain output from each hectare cultivated. But figure 3a suggests that in the United States, with its much larger land area, pressure to raise output per hectare was for a long time less.

One moral to draw from the diagram is that output figures for agriculture (or anything else) do not represent progress, even in a narrow technical sense, if they are taken out of context. We cannot simply conclude from the diagram that Dutch agriculture is more advanced than British, or that British farming is more efficient than American. Although data of this sort provide an insight into how techniques are improving, the diagrams that result need to be interpreted with care. If they are over-emphasized, they tend to foster a restricted perspective of progress which has justly been described as a 'one-dimensional' or a 'linear' view.[6]

Figure 3a also illustrates a more practical problem. Data available about past developments are often poor. Analysts are therefore tempted to present simplified diagrams which emphasize known long-term trends while omitting less clearly defined irregularities. That procedure can sometimes be justified, and the straight lines and angular bends in this particular diagram give fair warning of it. Yet on a previous occasion when the same graph was discussed, the author responsible was strangely deceived by its consistent upward slopes,

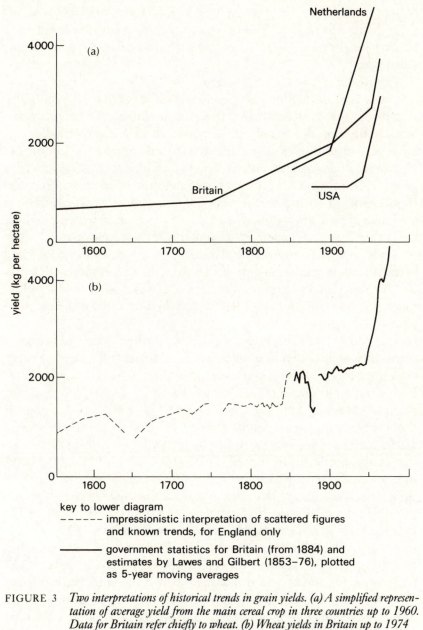

FIGURE 3   *Two interpretations of historical trends in grain yields. (a) A simplified representation of average yield from the main cereal crop in three countries up to 1960.
Data for Britain refer chiefly to wheat. (b) Wheat yields in Britain up to 1974*

Sources: (a) from J. R. Jensma, 'The quiet revolution in agriculture', *Progress: the Unilever
Quarterly*, **53** (4), 1969, p. 165; (b) plotted from data discussed by Hoskins and others
(see notes 8 and 9)

taking them at face value. He commented that in the United Kingdom, 'where records go back furthest, wheat yield has increased continually since 1350. Up to 1750, this increase was very gradual, amounting to about 1 kg a hectare a year. After that date, the annual increase became steadily greater.'[7]

This is an unusually clear example of the linear interpretation of progress, and of its flimsy empirical basis. One of the dangers in this instance is that it affects expectations of the future. An upward rising line like the ones illustrated, encourages us to think that recent dramatic increases in grain yields are securely based on several centuries of consistent improvement, and we then feel confident of maintaining this position in the future. However, another way of plotting this graph (figure 3b), in which all the major irregularities for which there is evidence are shown, reminds us that improvements in wheat yield have not really been achieved in a smoothly dependable manner. The present trend, based on chemical fertilizers and pesticides, on intensive mechanization, and on new strains of wheat, began only after 1945. It is not the outcome of long experience. Other periods of rapid development, for example in the 1840s, were based on quite different principles.

The presentation of figure 3b, with its dashed lines and breaks in continuity, is intended to show that most of the data for earlier periods are either inadequate or impressionistic. Authors who have discussed such data, including W. G. Hoskins[8] and Susan Fairlie,[9] point to long periods when very little sustained advance occurred; they also indicate times when crop yields fell as a result of climatic setbacks – around 1650, for example, and more briefly in the late 1870s. However, an especially significant feature of the history of wheat in England is the high level of output which was reached between the 1840s and 1870. Some statistics exaggerate this, but even the conservative estimates plotted in figure 3b show that wheat yields during the 1850s reached an average level which was barely exceeded for a century. This achievement arose from a system of agriculture known as high farming. It was based on a careful balance between animals and crops, on fertilizers such as guano, and on a fairly substantial investment in land drainage.

However, high farming also had a characteristic organizational aspect, reflecting its dependence on the rotation of livestock with crops. This required a complicated timetable for grazing different areas of land as well as for cultivation and sowing. Some tasks needed a very large labour force if they were to be carried out on schedule. Harvest was one, but so were some aspects of land preparation. Weeds such as twitch and couch-grass were controlled by repeatedly harrowing a field to bring the offending roots to the surface, and then employing an army of women and children to go over the land picking out the roots by hand.

The contrast between this system of practice and the pattern of advance after 1945 could hardly be greater. Now tractors enabled farmers to keep to the optimum schedule for sowing even when weather conditions were poor, and the use of herbicides enabled weed control to be effective without a massive labour force. Chemical fertilizers made possible the abandonment of complex rotations involving livestock, and thus in some ways made the organizational aspects of farming simpler.

Perhaps the most significant features of the diagram illustrating wheat yields are the two periods of rapid progress it illustrates, corresponding to the development of two contrasting forms of technology-practice – high farming and chemical farming. If the graph is to be reduced to a simple shape, that cannot be a smooth curve or rising line, but must include two near-vertical steps with a level area in between. Each step can be regarded as representing a distinct movement in agricultural practice.

Similar trends can be distinguished in other branches of technology. Later, we will see how the thermal efficiency of representative types of steam engine has improved from the time of the early Newcomen steam engine (first used in 1712) to the giant turbine units in modern power stations. Many commentators have used graphs to portray this sequence of advances, and as with wheat yields, the tendency among almost all of them has been to simplify the diagrams in order to show a consistent linear pattern of progress.[10] But again, there were periods of slow development and contrasting movements of more rapid advance when numerous innovations were 'clustered' together within a very short space of time. One of the best known of these movements occurred in the mining areas of Cornwall, where many steam engines were in use. There was a period of marked improvement in engine efficiency here during the 1820s and 1830s, following on from a slightly earlier cluster of innovations. Careful records were kept, and it is clear that performance did not improve altogether smoothly. Occasional brilliant achievements were not sustained, and phases of stagnation occurred when average performance hardly advanced at all. These fluctuations cannot be fully explained in purely technical terms.

However, Donald Cardwell suggests that the 'pace of advance generally', was set, 'not by the most brilliant and able engineers but by the capacity of the average individual – engineer or skilled mechanic – to master and use the improvements effectively'.[11]

As with the Indian hand-pumps quoted in the previous chapter, it is tempting for the engineer to think of steam engines only in terms of their mechanical working. The reality, of course, is that steam engines were dependent on men, not only to stoke the furnaces, but to adjust valves, oil bearings and carry out maintenance. So the performance figures one plots on graphs are not representative solely of the engines. Rather, they measure the performance of 'man-machine systems'; they measure the effectiveness of a particular way of practising technology.

The organization of work

The improvement of Cornish engines during the 1820s and 1830s, and the emergence of high farming in the 1840s, were both associated with the later stages of Britain's industrial revolution. The beginnings of that revolution are dated by some people to 1769, because in that year, two inventions were patented which are seen as especially crucial. One was a power-driven spinning machine devised by Richard Arkwright; the other was the first of James Watt's several improvements to the steam engine. This is a linear interpretation, however, emphasizing hardware rather than human activity, and presenting the industrial revolution primarily as a technical revolution. It is an interpretation that forms part of the conventional wisdom, and is alluded to regularly in speeches by politicians and industrialists, often with symbolic mention of Watt. One politician is quoted later (p. 26); another has said: 'the development of steam for the factory . . . produced a new economic system: capitalism'.[12]

Such opinions cannot be substantiated. The evidence is that the first factories of the industrial revolution, and the system of capitalism that went with them, did not depend on the steam engine at all. By the end of the 1740s there were several textile factories in Britain with power-driven machines, and more after 1769, but these used water-wheels or horses as their energy source, and it was not until 1783 that steam was first employed. In any case, most 'so-called factories were no more than glorified workshops', with machines 'powered by the men and women who worked them'.[13] Granted, the factory system could not have expanded very far without steam power. But primarily, the factory was an invention concerning the organization of work, with an earlier origin than most of the machines it contained.

Merchants trading in cotton and wool, yarn and cloth, wanted better control over production than they could achieve while spinners and weavers worked in their own homes. They believed that if they brought these people together in supervised workshops, they could stop embezzlement of materials, achieve more consistent quality,[14] and enforce longer working hours and a faster pace of work. Early writers on the factory system stress these organizational advantages. In 1835, one admirer of Richard Arkwright wrote that to invent a spinning machine was less remarkable than Arkwright's other achievement, 'To devise and administer a successful code of factory discipline'.[15] Rather less often, modern writers remind us that the essence of the early factories was discipline, and the opportunity which this gave entrepreneurs regarding the 'direction and coordination of labour'.[16]

There is a wider point to be made as well. Before the introduction of the factory system, hand-workers in cottage industry could set their own working hours and pace of work; before the agricultural enclosures which preceded the development of high farming, most people involved in agriculture had a similar freedom in their use of common land. But by 1800, individual workers were becoming employees of a factory-owner or farmer, and had to observe the working hours and procedures set by him.

Not only was this change in the way work was done central to early industry, but in many respects other aspects of technology lagged quite seriously behind. The machines in early factories were very simple and built mainly of wood. However, they worked just about well enough to demonstrate the potential of the factory, and the demand grew for better machines, made of more durable materials. By 1790, these were increasingly often driven by steam engines, which of course the British had pioneered. In some other respects, though, Britain was technically backward and needed to pick up many ideas from abroad. The inventors of early spinning machines had got ideas from Italian silk-reeling mills. After the 1790s, techniques such as chlorine bleaching and Jacquard's loom were borrowed from France, and the idea of calico printing from India. The first attempts to solve the problem of better factory construction depended on copying ideas used in building theatres in France; and to meet needs for improved transport, Englishmen studied French textbooks on bridge building and canal construction.[17]

However, these borrowed ideas were applied in Britain in an empirical, almost makeshift manner. There was little attempt to improve engineering theory or to teach it systematically in the way that was already done in France and later Germany. Indeed, technical education in Britain remained woefully inadequate throughout the nineteenth century.

Like all complex phenomena, the industrial revolution had multiple causes. Some were linked to banking and finance, some to the availability of material resources, some to population trends. Britain's essential contributions were related to all these, also to empirical skill, and above all to insights into the organization of work. But as science-based industries emerged later, Britain was less well equipped than Germany or France to take a lead.

Changes in the organization of work not only meant an enforced pace of work and fixed hours, but especially the division of labour. Complex tasks were broken down into a series of very simple operations, each of which was done by a separate worker. Wherever possible, machines or special tools were introduced to make these elementary operations even simpler, so that less skill was needed. The aim was to diminish the cost of the necessary labour, by 'substituting the industry of women and children for that of men; or that of ordinary labourers, for trained artisans'.[18] Thus work was both fragmented and deskilled.

The way in which deskilling is influenced by mechanization has often been discussed in relation to the development of machine tools, of which the lathe may be taken as an example. Early lathes simply held the material being worked, allowing it to be turned in a steady, controlled manner. The operator would manipulate the cutting tools by hand, and would turn the lathe manually, often by means of a treadle. As lathes developed, different parts of the task were taken over from the human operator. Power was applied to turn the lathe, and the treadle disappeared; devices were added first to hold and later to control the cutting tools. By analysing all the different aspects of a lathe operators' work, it is possible to identify in some detail the steps by which mechanical developments gradually took over more and more tasks, until today, numerically-controlled lathes are entirely automatic. Some authors have taken this analysis to the point where they can count each of a number of carefully defined steps in the development of the machine and use the resulting figure as an 'index of mechanization'. Ian Crockett[19] tied analysis of this sort to a chronology of lathe development, enabling another graph illustrative of technical progress to be produced (figure 4).

This diagram demonstrates very clearly how ambiguous linear views

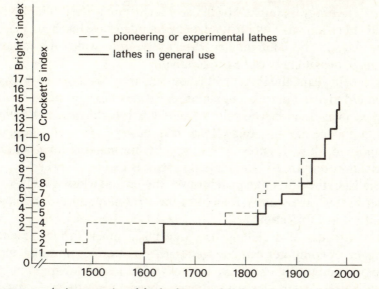

An interpretation of the development of the lathe since about 1400, based on the idea that an 'index of mechanization' can be used to measure the number of tasks carried out by the machine

> *Bright's index is discussed at length by Harry Braverman (see text), who suggests that lathe operators begin to experience adverse deskilling when the index reaches 5 or 6. Numerically controlled lathes have an index of 10 or more. Crockett's index was devised by Ian Crockett at the University of Manchester Institute of Science and Technology in 1971, and is based on the theory of kinematics. The original work remains unpublished.*

of progress can be, for it shows only the technical sophistication of the lathe, not its human significance. Moreover, the index of mechanization is in many ways arbitrary, and Crockett's version does not match precisely with the other index noted in figure 4.

From one point of view, we also need to notice that the early development of the lathe actually enlarged human capability. Without the application of power, and without devices to hold and control cutting tools, it would have been impossible to machine heavy metal products. Harry Braverman[20] also suggests that in the early stages, the greater speed and precision of work may have called for greater skills from operators. However, development did not stop there. As lathes became more nearly automatic they displaced not only muscular and manual skills, but the operator's judgment as well.

It is impossible to say precisely when lathe development passed from work enhancement to deskilling, though Braverman makes suggestions. But clearly, from the operator's point of view, the impression of consistent progress implied by figure 4 is extremely misleading. If one could check job satisfaction, it might be found to have reached a peak at some date in the later nineteenth century, after which the negative effects of deskilling became predominant.

In recent years, the division of labour and the deskilling of work have extended to many more occupations, aided very often by the computer. Mike Cooley describes how the engineering draughtsman in the 1930s was at the centre of design work in industry. 'He could design a component, draw it, stress it out, specify the material for it and the lubrication required.' Nowadays the work is divided between several specialists: the 'draughtsman draws, the metallurgist specifies the material, the stress analyst analyses the structure and the tribologist specifies the lubrication'.[21]

The process of deskilling has been taken much further recently, however. A computer can now generate the drawings on which many draughtsmen would once have had to work, and the designer himself, using a VDU as an electronic drawing board, can produce drawings and detailed designs much faster. But even the professional in charge of a computer design facility may find his work partly deskilled as systematized design procedures are programmed into the computer to form what Cooley calls an 'automated design manual'. Thus the work of the designer has been undergoing exactly the same process of deskilling as manual work. Sometimes he is reduced to making a series of routine choices between fixed alternatives, in which case 'his skill as a designer is not used, and decays'.

Obviously, computer-aided design has the potential to be used far more creatively. In more modest ways, word processors applied to office work present a similar choice between systems that are planned merely to speed up work and increase management control, and systems which enhance job interest. On the shop floor, when numerically-controlled machine tools are used, the advent of microprocessors provides an opportunity for building in 'more opportunity for shop-floor intervention to improve performance'. In each of these instances, though, managements are often reluctant to take advantage of the more creative option, 'because it would lessen *their* opportunities for control-ling output'.[22]

While it is acknowledged that much tedious work is displaced by this type of technology, claims are made about the new skilled jobs being created, which relate both to software and to the maintenance of equipment. However, the philosophy of maintenance has altered so that the work involved has also been deskilled. Braverman notes that even householders have observed a deterioration in the repair skills of the men who service their washing machines; modern equipment is designed as a system of standard modules which can be replaced without much knowledge.

Undoubtedly, computerization can help us cope with the complexity of the modern world and the pressure of resource shortages, but the problems associated with it should not be disguised. Microprocessors allow many kinds of equipment to be more compact and energy-efficient. Computers and modern communications may allow a trade in information to grow at a time when trade in material goods could become restricted. This would make knowledge more important as a resource in its own right; one consequence already evident is that computers change power relationships within firms and within the community as knowledge-based power is reduced for some people and increased for others. The result is that computerization is tending to 'strengthen rather than weaken . . . centralization and hierarchy' in modern organization.[23]

Harry Braverman draws an instructive comparison with the first industrial revolution. That was not primarily a technical revolution; there was no change in the nature of many processes, which were merely reorganized on the basis of the division of labour. Craft production was dismembered and subdivided so that it was no longer 'the province of any individual worker'. In the modern 'revolution' the whole system is transformed. New materials, techniques and machines are used in an effort 'to dissolve the labour process as a process conducted by the worker and reconstitute it as a process conducted by management'. The individual workman or operative is analysed almost as a piece of machinery; he or she is seen as a 'sensory device', linked to a 'computing mechanism' and 'mechanical linkages'. This, says Braverman, is what modern industry 'makes of humanity'; labour is 'used as an interchangeable part' and progress is seen as a matter of indefinitely increasing the number of tasks that can be carried out by machine. The final triumph is achieved when all the human components have been exchanged for mechanical or electronic ones.

Determinist deductions

Beyond value judgements about the desirability of all this, two contrasting beliefs about progress are evident. There is the linear view, well expressed by the graphs showing smooth, steady, upward development. But in sharp contrast, there is a view in which the context of innovation – including its organizational aspect – is more broadly considered.

Beliefs based on both approaches have some validity, but the linear view oversimplifies matters in a way which encourages a false optimism among those who approve the kind of progress it portrays, and a deep, despairing pessimism among others. The reason is that technical progress, as the linear view presents it, may seem inevitable and inescapable. There is a consistency in it which appears to imply an inflexible logic. When graphs conceal ambiguities and smooth out irregularities, they may seem to indicate an unvarying pattern in the unfolding of technological rationality – a pattern which is independent of the ups and downs of human affairs. This regularity, based on the internal logic of technology, has sometimes encouraged efforts to discover the 'laws' which supposedly govern progress. For example, one of two laws put forward by Jacques Ellul is that technical progress tends to act according to a geometrical progression because, 'the preceding technical situation alone is determinative'. He argues that, 'Technical progress today is no longer conditioned by anything other than its own calculus of efficiency',[24] and this leads him to a bleakly pessimistic view. Others have claimed that because technology 'carries its own culture with it', it also 'determines the ownership structure of industry'.[25]

All these views are variants of an attitude often referred to as technological determinism, which presents technical advance as a process of steady development dragging human society along in its train. Then many social problems are regarded as being due to 'culture lag', which arises when social norms and institutions fail to adapt to the latest developments in, say, automation or cable television.

This idea of technical advance as the leading edge of progress is widely held; it constitutes what some have called 'machine mysticism'. Thus we see ourselves as living in the computer age or the nuclear age which has succeeded the nineteenth-century age of steam. Each era is thought of in terms of its dominant technology, extending back to the early history of man. There, we think of development from stone tools to bronze ones and the later emergence of an iron age, as a logical technical progression bringing social evolution in its wake. And we think about each era in terms of the impact of technique on human affairs, rarely enquiring about the converse.

The Chicago World's Fair of 1933 was a particularly strong expression of this view of technology and progress in the way its exhibits were arranged and presented. The guidebook described how 'science discovers, genius invents, industry applies, and man adapts himself to, or is moulded by, new things'. It went on to assert that individuals, groups, 'entire races of men' are compelled to 'fall into step with . . . science and industry . . . Science finds – Industry applies – Man conforms.'

Such views are not often stated so starkly today, but they still seem to lie at the back of people's minds. Yet quite often, new patterns of organization had to be invented or evolved before innovations in technique could arise. The idea of television, for example, could hardly have arisen in a society without mass entertainment and organized news media. Thus Raymond Williams has described how radio and television evolved from an urbanized institutional background in which there was a growing need for communications of all kinds.[26] It is not sufficient to consider only the technical logic of the first photo-electric cells and cathode ray tubes, invented just before 1900. These, perhaps, were the beginnings of a capability to create television, but there had to be a conscious intention as well, and this was to a large extent 'an effect of a particular social order'.

In a similar way, it is possible to turn the conventional history of almost any invention on its head, and instead of showing how technical developments grew one on another, influencing social change, we may demonstrate how organizational development led to new technology. As already noted, instead of saying that James Watt's steam engine led to the industrial revolution, it is possible to argue that the prior development of factory organization gave Watt the opportunity to perfect his inventions.

As Williams says, technological determinism is untenable – but so is its complete opposite. Most inventions have been made with a specific social purpose in mind, but many also have an influence which nobody has expected or intended. The reality is perhaps easier to comprehend by thinking about the concept of technology-practice with its integral social components. Innovation may then be seen as the outcome of a cycle of mutual adjustments between social, cultural and technical factors. The cycle may begin with a technical idea, or a radical change in organization, but either way, there will be interaction with the other factors as the innovation comes to fruition.

This applies as much to the stone and bronze ages through which early man passed as to the industrial revolution. Explanations are insufficient if they focus only on the development of tools; there is also need to recognize the 'whole complex of mutually enhancing agencies arising from ecological circumstances . . . tool using and making, symbolic communication . . . group conduct'.[27]

So in the development of modern technology, it is not just the influence of tools and techniques on society that needs to be understood, but the whole array of 'mutually enhancing agencies' which has led to the spectacular advances of our own time. As another student of human evolution has put it, 'technology has always been with us. It is not something outside society, some external force by which we are pushed around . . . society and technology are . . . reflections of one another.'[28] Equally, it is a myth that cultural lag occurs in every community as people try to keep up with their progressive technology. In the interactions which take place between various aspects of human activity 'it is often technology that is lagging'.[29]

With such views so clearly expressed, why are conventional beliefs about the inevitability of technological progress and its leading role in social development still so widely held? The answer seems to be partly that the conventional beliefs serve a political purpose. When people think that the development of technology follows a smooth path of advance predetermined by the logic of science and technique, they are more willing to accept the advice of 'experts' and less likely to expect public participation in decisions about technology policy. Thus Leslie Sklair remarks that, 'the argument about the intrinsic dynamic of science and technology seems to me . . . the defence of those who find the idea of the democratization of science and technology unpalatable'.[30]

Such people argue that technical logic 'determines a unique progression from one stage of development to the next'. The implication is that although we may not like the idea of nuclear power, microelectronics, or heart transplant surgery, we have to solve the technical problems connected with these things if engineering and medicine are to develop. Such attitudes were reflected by the comments of a British government minister in December 1978, when he launched an initiative to promote the manufacture of 'silicon chips' and microprocessors. He remarked that we cannot 'stop technology'; it would have been no use trying 'to stop the steam engine, or . . . electric light', and it is no use now 'trying to stop the silicon chip revolution'.

That kind of statement is clearly designed to defuse political dissent – but it puts the issues the wrong way round. Not many people want to stop microelectronics, but they may want to state preferences about how it is used. The minister's remarks present a view of progress which implies only one dimension of choice: either you accept innovation unreservedly, or you opt out. Silicon microchips have potential for many kinds of development, though; it is choices between these that matter.

The development of the steam engine was by no means inevitable and 'unstoppable'. On the continent of Europe its adoption was halting and slow. Rapid development in Britain reflected the success of the new ways of organizing industry, and the freedom of mine owners and factory masters to pursue their ends without much social or political restraint. If the advance of microelectronics now seems inevitable, we ought to ask what kind of organizational pressure lies behind it, and what restraints might be appropriate in order to give proper effect to the wider public choice? We need beliefs about progress which help us recognize the real choices available. Existing views about how technology develops seem mostly to hinder perception of choice and allow experts and the industrialists they serve to get their own way.

Explanations of the rapid development of the steam engine in Britain and its slow progress on the continent include another point of modern concern. From 1712 until Watt extended the range of their application, most steam engines were used to pump water from coal mines. In any other situation, the early, inefficient engine used far too much fuel to be of much economic worth. And Britain was far in advance of most of Europe in mining coal, because deforestation had gone much further in Britain than elsewhere, and firewood was scarce and expensive. In much of the rest of Europe, as in America, wood fuel continued to be relatively abundant for another century; so there was less need to mine coal, less need for engines at mines, and little other opportunity for the economic application of steam. Thus there were strong ecological or environmental reasons for the adoption of steam power in Britain which were less important elsewhere. But there is no ecological determinism, and the environmental condition of Britain in 1712 did not dictate the development of the steam engine. A vigorous policy of reafforestation and fuel-wood coppicing could have solved the problem.

So also in our own immediate future, shortage of conventional energy will not eliminate choice; there is no determinism making it inescapably necessary to extend nuclear power systems, or to develop solar energy techniques. The issue, rather, is whether those who make decisions will prefer that we pay the environmental and social cost of one option or others – or whether we shall reduce these costs by conserving energy and by recycling materials. The choices available, in any case, are much wider than decisions about energy. As we have seen, the growing importance of knowledge as a resource means that industry may already be developing in directions that depend less on materials and energy.[31] Other options are possible, some relating to social policy (chapter 4), and others influenced by voluntary changes in lifestyle such as individuals have sometimes already adopted. The future is rich with choice, and the most balanced view of the environmental problem is not that it 'destroys the notion of . . . progress',[32] nor that it dictates the adoption of any particular technique, but that it requires, 'a new evaluation of what does and does not constitute "progress" '.[33]

Movements in progress

One way of rethinking our concept of progress may be to take an altogether broader view of the many factors which interact in 'mutually enhancing' ways at especially creative moments. At such times, the various technical, organizational and cultural workings of technology-practice seem all at once to start meshing together in new and more harmonious, effective ways. A new pattern emerges, and people experience a new awareness of practical possibility. The age of Columbus and the discovery of America was such a time. People had long realized that the earth is a sphere – Columbus did not have to teach them that. What happened was the dawning of recognition that this familiar academic fact had an unrealized practical potential.

Much the same can be said about the invention of the factory system. Many of the ideas on which it was based were commonplace. Italian merchants had operated the division of labour and a degree of mechanization in textile manufacture three or four centuries earlier, and the division of labour had been discussed in Britain since the 1650s. What happened at the start of the industrial revolution was a recognition of how to fit these ideas to economic opportunity in effective ways. And in the enthusiasms of the time, voiced particularly by Adam Smith, one may catch the same sense of a reaching toward a new and enormous potential that one feels with the age of Columbus.

There is clearly no determinism about these experiences of progress. A human, not a mechanistic process is at work, in which there is certainly an element of choice. But choice here is not the simple weighing of known options – it involves, rather, different ways of approaching the unknown. It is a decision between different attitudes of mind. We may cultivate an exploratory, open view of the world in which awareness can grow; or we can maintain a fixed, inflexible view in which new possibilities are not recognized. The magic of European culture from the age of Columbus until after the time of Watt was perhaps chiefly in its openness and growing awareness. The dominance more recently of linear views of progress, which restrict expectations to narrow patterns of development, may be symptomatic of how that openness has been lost.

Rather than speculate on such general matters, however, it is more realistic here to think again about the way improvements in agriculture and in steam engines have clustered together, leading to a step-wise pattern of advance rather than smoothly continuous progress. Each step, whether representing Cornish engines or modern farming, was characterized by specific organizational arrangements as well as by new techniques; it thus seemed right to describe these distinct phases of development as movements in technology-practice. What we can now add is that innovation is not simply the outcome of rational logic. It involves purpose and intention, and reflects awareness of possibility and economic opportunity. So in these minor technological movements, as well as in larger developments, there are crucial moments of recognition when a varied collection of different factors fit together and a new form of practice takes off.

Examination of diagrams, such as those presented in this chapter, enables a few of these moments of take-off to be identified. Figure 5, for example, illustrates the effects of several innovative movements in the use of steam power. The two upward steps in peformance associated with James Watt and with Cornish engine development are clearly seen.

In more recent times, the outstanding innovative movement involving steam power has been the development of electricity generation using steam turbines. This movement acquired a very particular self-awareness in Britain during the 1920s and 1930s when efficiency figures for individual power plant were regularly published by the Electricity Commissioners. Leslie Hannah[34] quotes the data and describes how engineers vied with each other to be at the top of the league table for their type of plant. The development of electricity supply was a movement in a much wider sense, though. It evoked considerable idealism about the creation of clean, pollution-free cities

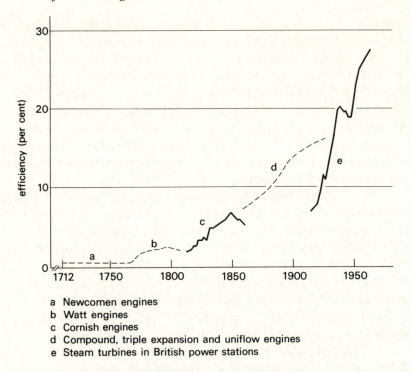

a Newcomen engines
b Watt engines
c Cornish engines
d Compound, triple expansion and uniflow engines
e Steam turbines in British power stations

Data have been recalculated to a common base, in which a full line is used only for consistently recorded series. Untypical efficiencies recorded during special engine tests are not shown.

FIGURE 5 *Average efficiencies of representative British steam engines and turbines (with boilers) from the invention of the Newcomen engine*
Sources: see chapter 2, note 34.

and factories. In England, it brought the vision of the garden city within reach. Women actively campaigned for it, seeing the use of electricity in the home as a step toward emancipation (see p. 106 below), and so did the Labour movement. It is ironic that in the 1980s, this progressive mood has largely dissipated, and the electricity industry is now the target of criticism about its over-ambitious expansion (p. 39 below), and its role as a source of chemical, nuclear and thermal pollution.

There is a further point to notice: innovative movements are not usually restricted to specialist fields. There were links between the achievements of Cornish engineers and the development of steam power for railways. There are strong and obvious links between modern agriculture, with its remarkable crop yields, and the chemical industries that produce fertilizers and pesticides. New electrical appliances were invented, and older industries were modernized as elec-tricity supplies developed.

Because of these links, a clustering of innovations in one branch of technology may lead to inventions in a wide variety of others; indeed, detailed analysis shows a distinct bunching of inventions at key points in the overall history of industry. This has led to the suggestion that we should not picture the last two centuries of technological progress as a smooth, continuous advance, but more as a succession of waves of innovation. Christopher Freeman puts this point by saying that inven-tion is not characterized by a 'linear trend', but by peaks and troughs, with peaks occurring as linked developments in 'new technology systems'.[35]

Thus when a large number of important developments occur within a couple of decades – as they did during the 1870s and 1880s with chemicals, steel, electricity and automobiles – then the economy may enter a phase of rapid expansion stimulated by several different tech-nologies simultaneously. Freeman points out, though, that a major economic upswing of this kind not only depends on technical innova-tion, but also, very often, on changes in organization; every new phase of economic expansion depends on 'a whole cluster of related . . . innovations and institutional changes'.

Freeman and his predecessors have meticulously charted the varying number of patents taken out from one year to the next in order to gain firm evidence for the clustering of inventions, and mainly agree with earlier work by Kuznets[36] about dates of the chief phases of technological development. This leads to a perspective in which four 'long waves' of industrialization are distinguished (and some shorter cycles also), the first being the classic industrial revolution in Britain. With each wave, inevitably, there is a time lag between inventions being patented, their coming into limited use, and their later impact on the economy; table 1 quotes two sets of dates to indicate this.

In a time of economic recession, such cyclical theories are under-standably more attractive than linear ones. They offer the hope that, after the cyclic downturn, there will eventually be recovery and renewed growth.

There is danger in this model, however, as much as in linear views. Any historical analysis which seeks to identify patterns and rhythms in development may tend to become deterministic. It may seem to imply that processes are at work which no human intervention can decisively alter.[37] Because there have been four long waves of industrialization,

TABLE 1   Long waves of industrialization

The four historical waves are, with minor modification, those identified by Christopher Freeman and his colleagues, and before them by Simon Kuznets and N. Kondratieff (see chapter 2, notes 35 and 36).

Dates for clustering of innovations	Key innovative technologies	Geographical base	Period of rapid economic growth
1st long wave			
1760s	textile manufacture	Britain	1780–1815
1770s	(also steam engines, figure 5)		
	chemistry, civil engineering	France	
2nd long wave			
1820s	railways, mechanical engineering	Britain, Europe	1840–70
3rd long wave			
1870s	chemistry, electricity,	Germany,	1890–1914
1880s	internal combustion engines	United States	
4th long wave			
1930s	electronics, aerospace,	United States	1945–70
1940s	chemistry (e.g. chemical farming, figure 3)		

POSITIVE PROSPECTS AND OPTIONS

5th long wave			
1970s	microelectronics, biotechnology	Japan, California	1985–?
Other waves?			
social development and improvement in quality of life with little economic growth	public health, nutrition	South Asia	
	renewable energy, conservation, agriculture, reafforestation	China, United States?	

some people are ready to expect a fifth. They see it as based on microelectronics and biotechnology, and centred on the 'Pacific rim' linking California, Japan and south-east Asia. But there is no inevitability about this. There are also those – economists as well as environmentalists – who think that industrialization is ending. They point out that all previous civilizations have had limited duration, and even if there is neither an environmental nor a nuclear disaster, we should expect that one day industrial civilization will go into decline.

There is also a third prospect, indicated speculatively on table 1 in order to stress that many options still lie open. This is the possibility of human development and progress involving technology, but independent of growth. A case-study illustrating such progress is presented in chapter 4, again with some graphical symbolism (figure 7).

Interpretation of the long waves of industrialization may easily become too narrow, because conventional analysis of economic and technical development must of necessity focus on facts and figures, graphs and statistics. Whatever individual authors intend, this inevitably gives little weight to human experience and choice, and leads to a mechanistic, if not a deterministic picture. To get beyond that point, we need to leave graphs and statistics behind and use other ideas. One social thinker who has done this offers an image which may help to show what innovative movements and waves may mean in human terms. With interests focused less on technology than on the broad social consequences of an energy crisis and an end to economic growth, he uses the analogy of people's habits in conversation. When one subject is exhausted, they talk about something else. Similarly, the goals of human development can change, be they technological or social, because, 'history proceeds by changing the subject rather than by progressing from one stage to the next'. Innovative movements and new waves of industrialization can be seen in human terms as changes of subject; and Ralf Dahrendorf, whose analogy this is, goes on to point out that what makes the difference is altered awareness:

> One day, people wake up to the experience that what was important yesterday, what preoccupied and divided them, no longer matters in the same way. We rub our eyes and discover that the way to solve the problem that kept us awake last night is not to do more or even to do better about it but to turn to something different which may be more relevant . . . we are in the process of one such historical change of subject . . .

In the advanced societies of the world, with their market economies, open societies and democratic politics, a dominant theme appears to be spent, the theme of progress in a certain, one-dimensional sense, of linear development, of the implicit and often explicit belief in the unlimited possibilities of quantitative expansion. The new theme which might take its place . . . is not a negation of growth . . . but what I shall call improvement, qualitative, rather than quantitative development.[38]

One does not need to agree in detail with the ideas which Dahrendorf goes on to develop in order to appreciate his starting point. The need is to keep open the possibility of waking up to the experience that there are new, radically different ways of dealing with economic problems, and that there are unexplored options for human benefit from technology. There is still the difficulty, however, that our habitual style of writing and analysis, whether in sociology, economics or technology, is itself basically linear. Its aim is usually to understand in depth rather than to broaden awareness. It is a style based on following logical connections, pusuing meticulous detail, and measuring whatever can be measured. Unless it is skilfully used, the very literary form of such discussion can itself trap one into a narrow, linear view.

This book is concerned more with awareness than analysis, and I have therefore needed to experiment with its style. Robert Pirsig described a motorcycle journey across America in one experiment of this sort. I am less venturesome, but sharing the view that history proceeds by changing the subject, I have adopted a style that does likewise. In this chapter, there have been shifts of scene between past and present – between agriculture and automation. In later chapters, there will also be abrupt changes in geographical subject, from Britain to ancient Greece and then to Africa, and from industrial North America to rural South Asia.

The result may be distractingly kaleidoscopic to readers expecting analytical argument. But elsewhere, I have written about the development of technical ideas in a more conventional way – and have fallen heavily into traps of linear interpretation.[39] Elsewhere, too, I have written about specialist technology keeping fairly strictly and logically within disciplinary boundaries. In this book, I seek a different and wider understanding, with the aim of holding together ideas that seem to be in conflict, and of enlarging awareness of goals – and potential – which such ideas may conceal.

3
The Culture of Expertise

Halfway technology

Technology is seen at its best, suggests Lewis Thomas, whose experience is chiefly of medical research, in the use of antibiotics, and in modern methods for immunization against diphtheria and childhood virus diseases. These techniques are decisively effective and relatively inexpensive; they may appropriately be thought of as 'real high technology'. In sharp contrast, organ transplants and cancer treatment by surgery and irradiation are simultaneously highly sophisticated and profoundly primitive, and Thomas describes them as 'halfway' technology.[1]

What makes the difference? Thomas thinks it is knowledge. When a problem is well understood, neat and cost-effective ways of tackling it are found. Halfway technology, he argues, is the result of trying to tackle problems that are only half understood. To find better solutions, we need more research.

On checking the relevance of these ideas outside medicine, however, we are likely to notice further reasons why expensive and elaborate procedures are adopted. Sometimes knowledge already available is not perceived and used. Even with regard to cancer, there are some who differ from Lewis Thomas about the need for research, and some who say that cancer is over-researched. Even those who believe in research as a means to understanding, do not always expect much from it in terms of treatment. In 1978, John Cairns warned that we should not expect any major breakthrough 'in the next ten or twenty years, or . . . for another century'.[2] That may seem to have changed in 1982, when he hailed a major discovery about genes in cancer as marking 'one of those wonderful moments in science when many lines of research come together and much that was obscure suddenly seems plain'. But important though this is for molecular biology, it can only slowly influence treatment, and the argument of Cairns' earlier book still applies – that the best way to reduce the toll taken by cancer is to apply the knowledge we already have. That it is not being very adequately applied is partly due to political factors, which Cairns mentions, but partly also to the way professionals are trained.

Doctors are oriented to curing disease, and they use knowledge selectively with that aim in view. Where there is no curative job to be done, as in preventive medicine or when patients are beyond cure, available knowledge may not be perceived. One doctor admitted privately that there are things which 'professionals are almost trained to ignore'. Similarly in other fields, experts are trained in tunnel vision. They learn to examine specialized aspects of problems with a concentrated attention that blinds them to other issues.[3] Food shortages and energy problems become narrowly technical questions, with many aspects of organization and use forgotten. The technology of a green revolution, for example, may be planned to increase food production without any clear idea of why food consumption is low (p. 56). Much the same is true of Britain's electricity industry, which has been 'built by institutions in no position to ask' detailed questions about consumption, or to query whether accepted technologies are 'the best to perform their end-use functions'.[4] In fact, the end use of much electricity is to provide heat for homes and offices. Yet electricity generating stations discard heat into the atmosphere via their cooling towers, and this could be used for heating buildings directly; in much of northern Europe, it is so used. But few British power stations have been sited or designed with any regard to the community's heat requirements, and 60 to 70 per cent of their energy throughput must be thrown away. Massive cooling towers, sculpturally beautiful though they may be, are symbols of this halfway approach.

In the water industry, engineers have similarly failed in the past to pay much attention to end-use functions. They thus failed to notice that there are many industries which can re-use water before discharging it to waste. More remarkably, they did not notice that between 15 and 25 per cent of water supplied in Britain is lost by leakage from pipe networks, which are sometimes very old and poorly maintained. While such matters were ignored, it appeared that a vast expansion of water supplies would be needed. Engineers busied themselves in seeking new sites for reservoirs and planning the construction of aqueducts and great new dams, notably at the enormous but under-used Kielder reservoir.

Such enthusiasm for enlarging supplies while neglecting questions of water use has been characterized by Paul Herrington as a 'supply fix' mentality.[5] He points out that attitudes are rapidly changing, but notes that for many years, rewards for senior management were related to construction projects and many costs were hidden. Moreover, training in the industry more or less ensured that the same approach would be reproduced from one generation to the next.

Big dams feeding leaking pipes – like electricity generating stations pumping heat into the atmosphere – illustrate clearly what 'halfway technology' is about. Those parts of the system on which the engineers have focused most attention are extremely impressive. Power station turbines are almost as efficient as the laws of thermodynamics allow. Concrete dams have become elegant and economical in design. But as Lewis Thomas says of halfway technology in medicine, from other points of view, some of these sophisticated techniques are profoundly primitive and disproportionately costly.

Related problems affect the international effort on drinking water and sanitation which the United Nations has called for during the 1980s. Supply-fix arguments are heard on all sides as figures are quoted to show how many people in the poorer parts of the world lack access to safe water supplies and adequate sanitation. These figures do not correlate well with the realities of hygiene – especially those that relate to sanitation – nor with users' perceptions of their needs. They measure the availability of equipment but not its relevance, nor its utilization and maintenance. They encourage the formulation of policy targets in terms of the supply of equipment – 9.5 million hand-pumps are said to be needed in the 1980s, and 13 million in the 1990s. Meanwhile, improvements in hygiene are defeated by over-emphasis on construction and neglect of maintenance planning, organized cleaning of facilities, and health education.[6] So again supply takes precedence over use.

Hygiene and maintenance are linked not only by neglect, but as related concepts of some intellectual interest for this book. From an engineering point of view, hygiene is a maintenance activity in which routine cleaning of equipment takes on a larger dimension connected with individual behaviour and personal cleanliness. Further, hygiene and maintenance, together with preventive medicine, challenge the usual focus of technology on problem-solving in that they are concerned with problem prevention. Engineers link these concepts when they talk about preventive maintenance and apply this idea to urban buses as well as to hand-pumps and sewers.[7] What it means is inspecting and servicing equipment and replacing vulnerable parts according to a carefully planned timetable, so that potential faults are corrected before breakdown can occur. By contrast, the repair of an installation after it has failed is a curative activity.

Yet just as doctors are oriented to curing rather than preventing disease, so the tunnel vision of other professionals often excludes or marginalizes maintenance issues. Pipes are left to leak in Britain and there are chronic hand-pump breakdowns in India (chapter 1). And this is not just because problem prevention is an awkward concept for those trained in problem solving. It is also because real high technology tends to be inconspicuous. Lewis Thomas cites immunization as an example. Maintenance is similar, and often involves routine, repetitive, even tedious work in addition. Yet in water and sewerage systems, which (with better nutrition) have made a bigger contribution to health in the modern world than medicine, performance depends fundamentally on maintenance. Indeed, the technician who does these tedious jobs well may indirectly save many lives, 'so that it can truthfully be said that a technician's value is greater than that of a doctor'. Yet his work goes unremarked, and he is poorly rewarded. With similar emphasis, hygiene has elsewhere been described as an 'invisible technology',[8] and to those who identify technology with hardware or suffer from tunnel vision, prevention, maintenance, organization and end-use are all invisible.

One spectacular example of the invisibility of organization is provided by the effort to control malaria in India during the 1950s and 1960s. In a massively remarkable campaign, the walls of every dwelling in the subcontinent were sprayed with DDT to kill mosquitoes that entered people's homes. This had an immediate impact on transmission of the disease, whose incidence fell to a very low level. But with this achieved, it was still necessary to sustain surveillance work, to nip any new outbreaks in the bud. This entailed checking on people with fevers by home visits and the routine examination of thousands of blood samples in hospital laboratories. But with malaria apparently defeated, it proved difficult to carry out these tedious maintenance tasks with conviction. When DDT made its first dramatic impact on the disease, 'people witnessing the miracle would be conditioned to thinking that the key to success was DDT, not organization'.[9] Then too much reliance was placed on the insecticide and too little on parallel mosquito control measures and on the organization necessary to sustain the programme. Once mosquitoes had acquired some immunity to the insecticides, the technical fix which the latter had provided began to fail. India experienced less than a million cases of malaria each year through the 1960s, but a rising trend took the annual total to 30 million in 1977.

Arms and over-prediction

In most industries, planning must be based on some kind of forecast of future demand. In the British water industry, we have seen, such planning has been biased by over-emphasis on supply. Herrington comments that 'everyone who has studied demand from any sort of independent standpoint must have his or her favourite example of overprediction', such as the forecast that in Birmingham, per capita water consumption, 'would increase by 200 per cent between 1965 and 2000'.

Much the same has been true of electricity, with the result that several countries found themselves with surplus generating capacity during the 1970s. Planners could not have been expected to foresee the economic crises of the period, but had they studied electricity use more adequately, they might have anticipated the impact of cheap natural gas, the possibility of market saturation, and the extensive scope for energy conservation.

In 1981, a British Parliamentary Select Committee which reviewed national policy for nuclear power discovered that even as it was taking evidence from the Central Electricity Generating Board (CEGB), that body was scaling down its forecasts for five years ahead by a factor of 7 per cent. No admission of this was made, however, leaving the Parliamentarians to comment that the credibility of much of 'the CEGB's subsequent evidence was undermined by this omission'. The science journal *Nature* commented that the CEGB had spoiled a good case for nuclear power by allowing their forecasts to be biased 'by arrogance . . . conscious laziness . . . and a failure to understand what they are for'.

In a similar way, highway engineers have sometimes justified the construction of new roads by producing inflated forecasts of likely traffic flows. In Britain prior to 1978, road planners tended to over-predict traffic, in certain cases significantly. A government-appointed inquiry found that forecasting methods were 'unconvincing . . . inherently unsatisfactory . . . contrary to common sense'.[10] As with criticism of the CEGB, there was an implication that biased projections were a result of petty dishonesty. But such suggestions in some ways miss the point. Professional culture places a high value on integrity, but in engineering especially, there is a long tradition of coping with uncertainty by incorporating enormous safety factors into estimates. When this leads to a bridge being over-designed in the interests of safety, such caution is admirable. With the rather different question of planning for the future, however, a similar approach is out of place. For one thing, professionals planning for their own field tend to focus selectively on its most successful sectors. Thus there is more meticulous planning for electricity than for energy as a whole, for highways rather than transport, and in the British health service during the 1950s, there was a tendency to overpredict need for institutional maternity care while underpredicting the need for care for the elderly. The problem is rarely a lack of integrity, or even conscious laziness, but the professional's sense of commitment to his own branch of expertise.

This mentality is disturbing enough in the context of roads or water supplies. But several analysts have identified a very similar approach to defence as one of the factors tending to accelerate the arms race. In the United States especially, inflated estimates and overprediction again appear as symptoms of the underlying attitude. In the 1950s, President Eisenhower's policies 'were always frustrated by those who consistently exaggerated the Soviet military threat'. Looking back from the 1970s, it could be seen that 'predictions made . . . over the past 20 years . . . have always been far-fetched', and based on 'phoney intelligence'.[11] Potential Soviet weapons production and actual output were not always distinguished, and Eisenhower's principal scientific adviser on defence, George Kistiakowsky, has described how the President was criticized 'for having allowed a . . . mythical missile gap to arise'. But it was a gap that never materialized.[12]

The result of such pressures was that one of the best opportunities there has been for limiting nuclear arms was lost. From 1955, the Soviet Union had shown some interest in an arms control agreement, and Khrushchev seemed genuinely concerned. One defence scientist who studied Khrushchev's speeches and writings in detail felt that this was 'a unique moment in history, when a man so open and so whimsical was in power in Russia. If we did not start quickly to negotiate with him about basic issues . . . the opportunity might be gone forever'.[13] The Eisenhower administration began to explore the possibility for a treaty banning nuclear test explosions, but this provoked 'disabling hostility from the British', who needed more tests to complete their 'independent' deterrent. More unrelenting, however, was hostility to a test ban from defence scientists in the United States whose research would be curtailed if a ban was enforced. Edward Teller was leader of this group, and some scientists who supported him felt that their motives were based on 'peaceful and pure'[14] research interests in the physics of nuclear fusion, and in nuclear-powered spacecraft. But there were certainly political motives too; their lobbying against the test ban was very determined, and the treaty was not signed and ratified until 1963 when Kennedy was president. Even after that delay, the defence scientists' opposition was only defused by restricting the treaty so that underground testing could continue.

Solly Zuckerman, a former scientific adviser on defence to British governments, has said that during the test ban treaty negotiations, the US government was 'at the mercy of the technical judgements of men who were concerned only with their departmental interests'. In 1976 and 1977, when a more comprehensive test ban treaty was within reach, American proposals for a seven-year ban were cut to five, then three years, and then dropped under pressure from the same professional interest groups. In 1982, the Soviet Union seemed to be indicating continued interest in such an agreement.[15] Indeed, a comprehensive nuclear test ban – with a ban on test missile firings also – is arguably the most urgently needed among those arms control measures that seem practicable, because it would set limits to armaments research. It would thus curb the tendency of the weapons technologists to escape political control.

Proposals current in 1982 to equip western armies for a 'limited' nuclear war in Europe seem to involve some parallel issues. Once again, Edward Teller and other defence scientists appear to have pressed strongly for the adoption of new weapons, including the so-called neutron bomb or artillery shell. Once again, a supply-fix mentality has been evident as the scientists argued for the weapon out of technical interest in its development and manufacture, and not because of any end-use demand. Indeed, military men who would be in charge of any end use do not universally welcome this new technology. Equally, clear statements about deployment of the weapons in Europe have come from parts of the British military establishment.[16] The late Earl Mountbatten, speaking in 1979, expressed despair at American resistance to arms control. 'What can their motives be?', he asked. 'As a military man . . . I say in all sincerity that the nuclear arms race has no military purpose. Wars cannot be fought with nuclear weapons.'[17]

Mountbatten was, of course, right. The predominant function of the arms race is not just military, but is also concerned with sustaining certain lines of technology, research and industrial development. But in order to persuade the nation to pay for such development, the defence scientists have to make the best of what information they can find about possible threats. One cannot help feeling that if Soviet Russia had not existed, it would certainly have been invented. As it is, hawkish speeches and arms build-ups in the West may have caused Russia to spend more on its own defence than its government wished, under pressure from its own defence scientists; and in these ways, the predictions of experts on both sides tend to be partly self-fulfilling.

The arguments in favour of planning for limited nuclear war again follow the supply-fix pattern, but now it is not even necessary to produce biased forecasts – current facts will do. There is certainly an imbalance in Europe between the Soviet bloc's 20,000 tanks (in 1981) and NATO's 7,000. If nothing is said about the better quality of the western tanks or about NATO's anti-tank defences, it is possible to envisage the whole of Europe being quickly over-run. Tactical nuclear weapons can then be presented as the only way to stop this. Indeed, this very argument was repeated by the Pentagon in July 1982 to justify plans to manufacture up to 3,000 of the new W82 neutron shells, in addition to an earlier type already in production.

Yet the great weight of informed opinion is that any supposed tank threat could be more effectively and less dangerously contained by conventional weapons. The American supreme commander in Europe, General Bernard Rogers,[18] said as much in September 1982. Others point out that new conventional technology allows clusters of anti-tank bomblets to do the job of a neutron bomb with less risk.[19] And Field Marshal Michael Carver points to the potential of some very neat technology that could be employed without resort to nuclear options. Conventional anti-tank weapons 'can now be produced that do not depend upon kinetic energy to penetrate armour'. That means that they can be very light – even portable by a man – and relatively inexpensive, especially regarding transport. Moreover, elaborate skills are not required, and reservists could use the system. Thus large numbers could be deployed against any supposed tank threat, but being defensive weapons, and of little use in offense, their deployment would not heighten international tension.[20] NATO's concept 'that it can avert conventional defeat by initiating nuclear war' is therefore not just folly, but as Carver asserts, it is also unnecessary.

Why, then, are neutron bombs being built? Why are nuclear arsenals still being expanded? Freeman Dyson, professor of physics at Princeton, believes that 'The intellectual arrogance of my profession must take a large share of the blame.' Conventional weapons, especially of the type that are defensive rather than offensive, 'do not spring like the hydrogen bomb from the brain of brilliant professors of physics', but are 'developed laboriously by teams of engineers'.[21] Professors have prestige and influence; engineers who do painstaking work have almost none. The engineers' work may be real high technology; it may require patience, organization and attention to detail, and thus may involve the values and attitudes we saw characterized by maintenance work in other technologies. But as Dyson concludes, defence is 'not technically sweet'.

Solly Zuckerman has said very similar things in a formidable stream of publications which categorize weapons scientists as 'alchemists of the arms race'. He asserts that the pace and nature of the arms race is determined, 'not by . . . governments but by a community of . . . scientists and engineers – among whom I have to count myself'. Technologists have made the world more dangerous by doing what they see as their job. To stop the arms race will thus require a new approach to the control of technology. It will require politicians to devise a control of research and development such as has never so far existed;[22] it will also require new attitudes and values within the professional culture itself so that technologists cease to see their job in quite these terms.

Multiple-cause problems

One implication of these comments by Dyson and Zuckerman is that the arms race is not only an issue of concern in itself, but is also partly a symptom of problems lying deeper in the professional culture of technology. Other major issues – resources and the environment, food and population – may be similarly approached. Thus the central theme of this book is not the specific environmental and military dangers about which so many others have warned, but the cultural problem that relates to all of them.

This is partly the problem that as knowledge increases, our appreciation of what we know seems to become more and more one-sided, so that we accept the development of costly and dangerous halfway technologies and regard them as progress. Often, as we have seen, this bias arises from thinking only about the supply of a commodity and not about the human aspects of its use. But another, related bias, arises from the habit of presenting complex problems as if they had single causes and therefore simple solutions.

For example, if my comments in previous paragraphs were to suggest that the sole cause of the arms race has been the lobbying of weapons scientists, that might lead to the conclusion that there is a simple solution available – lock up the experts. The influence which some scientists have been able to exert seems an especially significant illustration of what professional culture may involve. But it would be foolish indeed to represent this as the single cause of a very complex train of events. At least half a dozen other factors may also be regarded as contributory causes of the arms race; any effective arms control movement needs to tackle several, if not all of these simultaneously (table 2).

Political and industrial pressures, personal values and professional culture may all foster the temptation to look for a simple solution to any worrying problem, often in the form of a technical fix. For example, smog forms in some cities under certain weather conditions, and pollutants from automobile exhausts are a causative factor. A technical fix, once suggested by a Philadelphia chemist and backed by a local chemical firm, could be to spray the organic compound DEHA into the air when smog is likely to form. This could certainly stop the smog but at the cost of adding an additional pollutant to the air, with possible health impacts. Advocates of DEHA spraying have argued that the alternative strategy based on modifying automobile engines also produces new pollutants.[23] We ought to notice, though, that both technical fixes distract attention from social issues which ought to be considered – the working hours that create commuter traffic congestion and the habitual use of automobiles rather than public transport.

Also noteworthy in this episode is the way each professional interprets the problem according to his own specific type of expertise. The chemist studies organic molecules, the automotive engineer redesigns vehicles, and the highway planner looks for ways to reduce congestion. Yet situations of this sort are usually best tackled by a combination of measures drawing on several kinds of expertise in a co-ordinated way. Thus, table 2 illustrates how any effective campaign to stop the arms race will require a range of counter-measures drawing on different

TABLE 2   Some contributory causes of the nuclear arms race

Causative factors	Possible counter-measures
International	
1 International tension due to overt action, or miscalculation	
2 Distrust due to poor communications, and past experience	Increase exchange of information; free scientific exchanges and travel; 'confidence-building' measures as under the Helsinki agreement (1975)
3 Ideological competitiveness; superpowers' desire for dominance	
4 Technical competitiveness	Divert rivalry from military projects to space exploration, etc.
5 Destabilization of the balance of power by innovation in weapons or defence	Agree new test bans to slow innovation; share more technical information
6 Proliferation of nuclear weapons among other states	Strengthen and enforce the Non-proliferation Treaty (agreed 1968; ratified 1970)
National	
1 Industrial pressures (Eisenhower, 1961, warned of 'the military-industrial complex')	
2 Use of the arms industry as an economic regulator, to stimulate growth	Trade-union pressure for 'peace conversion' in industry
3 Lobbying by weapons scientists (Eisenhower warned of 'a scientific-technological elite')	Counter-pressure from concerned scientists and the peace movement; agree new test bans to contain research
4 'Phoney intelligence' (Zuckerman)	Promote independent intelligence and public interest research groups, e.g. SIPRI (Stockholm International Peace Research Institute)
5 Inter-service rivalry	
6 Inadequacy of NATO conventional weapons	Improve conventional defensive capability, e.g. anti-tank weapons

sections of the community – not only politicians, but concerned scientists, trade unionists and others.

The same is true of many other problems which have multiple causes. The effort to control malaria in tropical countries during the 1950s and 1960s failed not only for the reasons already mentioned, but also because of over-reliance on a single technique – insecticide spraying. As with the proposal to spray a chemical to control smog, this dealt with the problem by introducing a new pollutant. But with malaria incidence now increasing, and with greater awareness of the dangers and difficulties, attention is being given to a wider range of strategies, including medical measures directed against the malaria parasite as well as a more broadly-based attack on the mosquitoes that carry it. One programme in Sri Lanka, planned for the years 1982–85, seeks to organize village people for a variety of jobs ranging from filling in puddles where mosquitoes may breed to a limited and selective use of insecticides.

The temptation to look for simple solutions has also had a distorting effect on attitudes to other diseases. For a long time, people seemed to think that cancer should be tackled by looking for an effective cure, but that is now recognized as too narrow. With greater interest in prevention, though, there is now the difficulty that every expert has his or her 'one right answer' for tackling that, and the idea of using a combination of measures that could complement one another receives little attention. There are differences between how specialists perceive the causes of cancer; there are also differences in values surrounding the question of where responsibility for prevention should lie – with the individual, with industry or with government.

Awareness of the important role of smoking in lung cancer has led to emphasis on the individual's decision not to smoke, and the same approach to prevention can be extended to the various forms of cancer where diet, alcohol and sexual behaviour are relevant factors. Regarding diet, for example, high intakes of fat combined with insufficient fibre seem to be associated with cancer of the colon and breast cancer. Richard Peto, a prominent British researcher, believes that the 'fat-associated and smoking-derived cancers collectively account for more than half of all cancer deaths'.[24] Emphasis on this view – for which there is increasingly good evidence – can lead to the conclusion that cancer prevention depends on the voluntary adoption by individuals of a fairly abstemious lifestyle.

Critics of this voluntarist approach claim that its advocates in the medical profession are seeking to blame the patient and simultaneously to cover up for industries which are responsible for a wide range of occupational cancers. There is a recognized occupational risk relating to stomach cancers and lung cancer for many groups of manual workers, especially those working in dusty conditions (miners, building workers), and those whose jobs involve organic chemicals (painters, chemical process workers, seamen on tankers or handling tarred ropes and nets). There are other forms of cancer where the possible influence of occupational factors has been largely ignored, however, including cancer of the cervix. Here, Jean Robinson points out that the textile industry (especially spinning) may involve a risk factor, and shows that apart from a woman's own occupation, her husband's job may be relevant.[25]

While voluntarists note that cancer of the cervix correlates with virus infections spread by sexual promiscuity, there is an environmentalist point of view which relates it to occupation, especially in conditions where housing is poor, hygiene is difficult, and dust from a man's work is transferred to his wife during coitus.

Disputes between voluntarists and environmentalists have become very heated at times, erupting forcibly in the columns of *Nature*[26] during 1980–81, and the case for an effective campaign for the prevention of cancer may well have suffered. As in supply-fix arguments, figures are massaged to suit all points of view. The experts say that only 5 per cent of cancer deaths in Britain are due to occupational factors, while environmentalists estimate that the proportion is between 20 and 40 per cent. Confusion is created by thinking in terms of simple explanations based on single causes. Smoking and industrial pollutants are known to interact, and both could be implicated in many cases of lung cancer. While it may be true of an industrial community that smoking is a factor in 80 per cent of all lung cancer deaths, it may simultaneously be true that there is an occupational or pollution factor in 30 per cent of deaths, and that several other contributory causes are important as well. In other words, the arguments of the two sides are not mutually exclusive, and the percentages must usually add to more than 100. But Richard Peto is surely right to assert that no other industry 'kills people on anything like the scale that the tobacco industry does'. Its current sales drive in the Third World, 'if successful, will kill millions'.

In much of this argument, it is professional physicians and epidemiologists who have provided most of the evidence supporting voluntarist attitudes to cancer prevention while their environmentalist critics are often lay people – trade unionists appalled at complacent attitudes to occupational illness, and women annoyed by the way doctors seem consistently to misrepresent women's diseases. Some lay comment may well be exaggerated, but professional tunnel vision has restricted the investigations made by experts. For the specialist, it may be technically more interesting to note the relationship of cancer of the cervix with the herpes simplex type 2 virus than to study recorded information about the work of women and their husbands. Jean Robinson points out evidence on the latter which has not been followed up; she rightly comments that there is likelihood of 'real injustice to women whose deaths are at present attributed to their own promiscuity where other possible causes have not been examined'. Earlier in this chapter, we saw how chemists perceive exhaust pollution as a chemical problem and forget its connection with the way people live; we saw how water engineers have sometimes failed to study how water is used; it is not surprising, then, that medical experts have forgotten to look at data on the occupational aspect of an illness when there are more specialist medical aspects to study.

Mapping areas of misperception

The restricted style of thinking exhibited by many professionals is not usually politically motivated in the way that environmentalist commentators on cancer sometimes suggest. More commonly, it is related to the intellectual culture of technology, and to the habit of identifying technology-practice with its strictly technical aspects. This point was made in chapter 1 (p. 6) with the aid of a simple triangular map of technology-practice, and we can now usefully extend this map to illustrate how wide a range of problems is consistently misperceived.

The technical aspects of technology-practice most readily identified are machines and other hardware, chemicals and drugs, liveware, specialized techniques, and scientific knowledge; these may be found on the bottom half of the map (figure 6).

But technology also implies organization, including most obviously the organization of industry, of people's daily work, and of the technical professions. This can also be shown schematically on the map. However, servicing and maintenance activities cannot be so easily represented. Sometimes, the servicing of products (such as automobiles) is well catered for by the industries that produce them. Spare parts are

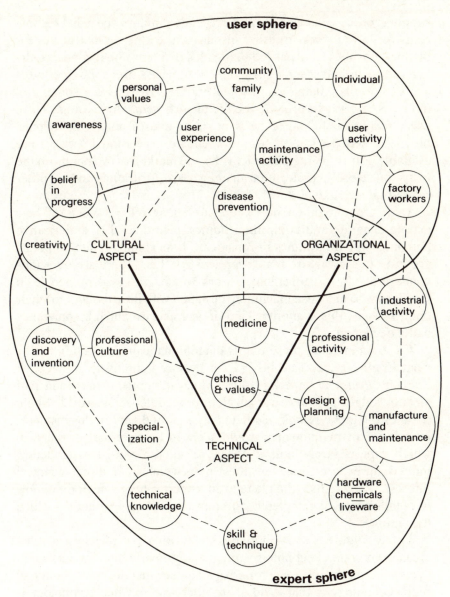

FIGURE 6   *An outline map of technology-practice developed from figure 1*

Users *are all those outside industry who operate equipment; who consume energy, food and water, and who make use of medical services. The* user sphere *indicates the main scope of users' organization and experience; some overlap with the* expert sphere *is indicated.*

supplied on a regular basis, and the trade employs specialist technicians to do the work. In other instances, the attitude that makes for planned obsolescence has led to the idea that household goods, clothing, radio sets and other items should be replaced after a short period of use rather than being maintained and repaired. Users sometimes improvise repairs to prolong the life of such items. Increasingly, also, users tackle maintenance tasks on their houses and vehicles on a do-it-yourself basis, even where adequate professional services are available. Thus maintenance activities are marked at two points on the map – one closely connected with industrial organization, and the other related to user activities.

Apart from maintenance, there are many other aspects of how people decorate and equip their homes, plan their diet, and arrange personal hygiene routines to consider. These and many similar points are indicated on figure 6 as 'user activity'. Then there are questions concerning the organization of work which are faced by people in factories who must use machinery designed with little thought for their point of view; this is another kind of user activity which is sometimes badly neglected.

To summarize all these points diagrammatically is to do in reality what Francis Bacon said he had done, more figuratively, in his book *The Advancement of Learning* (1605), which attempted to make a map of all knowledge, 'as it were a small globe of the intellectual world'. Some parts of the globe remain inadequately explored, and in this instance, knowledge of maintenance would seem to be on a part of the map which experts rarely look at. We have noted a consistent tendency, indeed, for experts to see only those parts which are of direct technical interest. They reduce the globe to an 'expert sphere' which they know in detail, leaving a completely different view – a 'user sphere' – which they ignore.

Thus on figure 6, two boundaries are drawn to encircle these distinct areas of interest. Technology-practice, if properly thought out, ought to consider both, and we may perhaps now see that the point about real high technology is that it does. In other words, when technology is really effective, this is usually because attention has been paid to maintenance and use of equipment, to users' or workers' or patients' knowledge and experience, to personal and social values, to government regulation of industry aimed at protecting health, and equally to the responsibilities of individuals for their own health.

By contrast, halfway technology is developed when professionals try to work in a self-sufficient way within the expert sphere. Equipment becomes over-elaborate and costly, and chemical sprays (for example) are developed because that is where the focus of the expert sphere lies. Industrial profit margins benefit from such bias in expert culture, and professionals who think this way get every encouragement. But as problem-solving based on expert knowledge and the products of industry presses forward, opportunities to prevent problems which depend on co-operation with users are missed; preventive medicine, hygiene and maintenance are neglected.

Two questions arise from this. One is whether the co-ordination between expert and user interests that seems necessary could be achieved if lay people were more actively involved in planning, design, policy-making and other areas which have often been of exclusive professional concern. I shall return to this question in chapter 8.

Another question, however, concerns what professionals themselves might do, or are already doing, to broaden their approach. Figure 6 itself might seem to imply one kind of answer, for this diagram is a form of 'systems map', and may seem to point to the application of systems theory as a way of avoiding some of the blind spots and misperceptions of professional thinking. Soft systems theory, for which one may see precedents as early as 1605 in the book by Bacon quoted above, is usually limited to exploring problems and taking soundings. It can be very valuable in a qualitative way to complement narrower, more analytical thinking in technology, rather as natural history complements more formal scientific method in biology: I habitually use its diagrammatic procedures. They are particularly helpful, for example, in clarifying the multiple causes of a complex phenomenon.

Similarly, Michael Collinson describes a procedure for developing a systems perspective in agriculture through teamwork involving both farmers and experts. The method brings the user sphere and the expert sphere closer together, and succeeds in this partly because it deliberately stops short of 'a detailed manipulation of numbers in a modelling format'.[27] Systems theory of the latter kind, usually worked up with the aid of a computer, sometimes serves only to lend a bogus air of precision to a basically imprecise approach. It even obscures the qualitative insights which are the chief value of the systems approach; and it provides experts with new techniques for mystifying and manipulating users. In this guise, systems theory finds its chief application as the 'ideology of bureaucratic planners and centralizers'.[28]

Another way in which professionals may sometimes attempt to

broaden their approach and avoid some of the more obvious blind spots is through interdisciplinary teamwork with experts from other fields. A Canadian study which indicates some of the problems relates to engineers and public health officers working in Vancouver and its hinterland. Both groups were concerned with river pollution, but it is clear that they perceived the same problems very differently. The engineers mostly favoured technical fixes for pollution, such as the construction of water treatment works, or diversions to increase river flow. However, the public health officials proposed to tackle pollution by negotiation with the firms discharging effluent, and if that failed, by legal action, and were altogether sceptical of the sufficiency of purely technical solutions.

Neither the engineers nor the public health officials collaborated regularly with members of other professions. They wished 'to retain complete jurisdiction' in their own fields, which they treated as 'closed systems'. They considered that their work was in the public interest, but regarded consultation with the public as 'either unnecessary or potentially harmful'.[29] Thus not only was the user sphere kept separate from the expert sphere, but the latter was further subdivided into smaller, self-contained specialisms. As the author of the Canadian study commented, unless 'experts broaden their views and integrate their activities, they may contribute more to the promotion of the environmental crisis than its solution'. In particular, they may concentrate increasingly on the creation of halfway technologies whose elaboration and relative inefficiency entails disproportionate environmental impacts or unforeseen side-effects.

How are these restrictions in professional life to be overcome? One group of doctors active in the 1940s called themselves humanists and were enthusiasts for what they called social medicine. They strongly emphasized a preventive approach to disease, and recognized that this meant breaking out of the conventional boundaries of medical science and moving into what I have called the user sphere. They needed to study conditions of life, 'in the home, the mine, the factory, the shop, at sea, or on the land'. Writing about this in the aftermath of the Second World War, when tuberculosis was still a common disease in Britain, and when malnutrition among the unemployed of the 1930s was still remembered, John Ryle wrote eloquently about the humanistic objectives of social medicine, and called for a 'crusade' to give due priority to the 'physical, mental, and moral health of people'.[30]

Although such attitudes still belong only to a minority, one finds occasional individuals today with the same social conscience and the same willingness to break through professional barriers that prevent real health problems from being adequately tackled. One such is a doctor from Kansas named Carroll Behrhorst, who went to Guatemala and opened a clinic in a rural area in 1962. Among the patients he saw, the majority had illnesses that stemmed from malnutrition. He realized that treating these people in his clinic only gave them a short-term remedy; as a medical professional, he had no self-sufficient answer, and no adequate fix. Thus he felt impelled to cross a professional boundary. His staff 'began to teach better farming methods'. They lent money 'to 25 families to raise chickens and produce eggs'.[31] Between 1963 and 1972, fifty argicultural extension workers were trained and a movement began which eventually helped several hundred farmers, allowing many of them to double or treble their crop yields. As more food became available, nurses from the clinic began teaching nutrition; people's diets slowly improved; malnutrition declined and tuberculosis disappeared.

There have been several other instances of hospital or clinic staffs who were confronted with malnutrition and initiated agricultural or vegetable growing projects. It has been possible at times to find medical personnel in Bangladesh and parts of Africa starting their working day with an hour's hoeing. Such activities are a very valuable gesture in a world of excessively rigid specialism. But even growing more food may be only half the answer when malnutrition is the result of poverty, and of the politics out of which poverty grows. I was made keenly aware of this when trying to collect information about health programmes in South Africa which promote vegetable growing; one of the most informative reports cannot be quoted, without risk to its author, because of a government banning order.

In many other instances, when we start looking at problems whole instead of just the technical detail, most of what we see is poverty. Much illness – malaria and often tuberculosis – is associated with poor housing. So is cancer of the cervix, which is more common in Bangladesh than Britain, more common in northern England than the south, and more common among American blacks and hispanics than among whites. Thus control of the disease by screening – available more readily to the better off – has been described as 'a betrayal of women of the Third World and our own poor'. Smoking (and hence lung cancer) is also more common among the poor, John Cairns points out, possibly because life offers fewer other pleasures.[32]

In Britain in the 1940s, John Ryle resisted the arguments of some reformers who wished to attribute all major disease to economic deprivation. As he rightly saw, there are always multiple causes. Yet he thought it right to leave clinical medicine himself in order to undertake research that was criticized as a 'social science' involving 'advocacy of social or other measures'.[33] He equally felt it the duty of the medical profession to 'do everything in our power to amend the graver inequalities in respect . . . of life and health', and questioned what this implied in relation to party politics.

What matters, he concluded, is not that all doctors should take to politics, but that they should be aware of the political, social and economic dimensions of the problems they face, as well as of the potential contributions of other professionals and also lay people when these problems are tackled. Such awareness can only come, to engineering as well as medicine, with changes in education and reforms in professional life. We need an atmosphere in which wide-ranging, interdisciplinary work or political involvement is not regarded as unprofessional; we need education which encourages the proper exploration of situations before there is a rush to problem-solving; we need to break down tunnel vision. Given these conditions, we would less often find potentially beneficial technology turning into distorted, damaging fixes.

4
Beliefs about Resources

Food and energy

The anthropologist Mary Douglas has described two neighbouring communities in Africa which live in the same climate but experience it in different ways. [1] What to one group seems the hot season is cool to the other. Beliefs about the environment are like that, she says; they relate to social systems and values, not only to facts, and in the western world, as in Africa, one can confront people with the same facts about the environment and find them coming to diametrically opposed conclusions. Economists and scientists, especially, are often found to interpret the same information about resources or pollution in different ways. Similarly, it was noted in the previous chapter that engineers and public health officials interpret river pollution differently, and that divergent views are held by informed people about the causes of cancer. Experts from different backgrounds are notoriously in disagreement about many other technical questions also – the nutritional requirements of the human body, for example, or the economics and safety of nuclear reactors.

Mary Douglas explains these differences by pointing out that we all need to find coherence and regularity in experience; we need a frame of reference or world view by which to order our perceptions. To achieve this, we must inevitably allow 'the destruction of some information for the sake of the more regular processing of the rest'. This destruction of information or backgrounding of matters we do not wish to think about may be done in several ways, for example, by rejecting ideas that do not fit into a linear view of progress, or by discussing the safety aspects of VDU screens rather than more serious worries about their social impact.

Sometimes this process can lead to a form of double-think. We are mostly well aware of the advertising, the marketing techniques, and the planned obsolescence by which manufacturers of domestic consumer goods, clothes and automobiles seek to promote sales. Yet we do most of our shopping in precisely the way that the manufacturers want, pushing our doubts into the background. Through similar backgrounding habits, professionals of unimpeachable integrity produce biased projections or preside over organized waste in the electricity and water industries without noticing it. All of us habitually do this, because otherwise the world would seem too complex to deal with. We accept a distorted world view in order to make the more immediate parts of our experience manageable.

One distortion which has become part of the conventional western world view is a caricature of the Third World and its myriad 'starving people'. Some organizations estimate that 1,200 million people are undernourished; others say 400 million; a few experts say less than 100 million. The figures are almost meaningless because of confusion about what they are meant to measure. Do they measure hunger, or rather poverty? Do they measure manifest symptoms of malnutrition or perhaps chiefly a risk of under-nutrition? And what criteria define under-nutrition anyway? Ambiguities about all these things give ample scope for the distressing but unmeasurable reality of poverty to be manipulated by cynics who would rather do nothing about it – or by powerful interests with fertilizer or farm machinery to sell, or surplus American wheat to dispose of. Some of the latter may want to believe in food shortages where none exist in order to justify their export drives, and their 'aid'.

Many people have become poorer precisely because of such exports – because the green revolution promoted by the West as a technical fix for a misperceived problem has deprived people of land and employment, and made many of them *more* hungry than before. Chronic malnutrition is certainly widespread, but it is not caused by single physical factors such as a specific scarcity of food. Rather it is a problem 'having multiple causes, many of which are closely linked to . . . conditions of inequality of resources, of poverty, and of social discrimination'.[2]

As to famine, studies in India, Nigeria and elsewhere have shown that its frequency and intensity have varied historically according to changes in the economic regime – not with the frequency of poor harvests alone. Under some economic conditions, for example, food prices are pushed up beyond what the poor can afford by speculative buying on the part of traders; and where farm output is depressed, this may be because prices paid to farmers are too low to encourage them to produce. 'The connections that do exist between . . . food scarcity and the environment are mediated by the political and economic relations of a society.'[3] Famine, like recession, is largely man-made.

Such facts are known, and in recent years some countries have made appropriate reforms in food marketing. Yet we persist in thinking about hunger as if there were an absolute shortage of food, and as if the problem of matching food supplies to a growing population were a purely technical one, to be solved by the new crops and fertilizers of a green revolution, supplemented in times of difficulty by food aid from Europe and North America.

The way in which westerners are willing to destroy information in order to maintain this belief is illustrated by the way in which food aid is sometimes dispatched even in the face of clear evidence that it is not needed. A striking instance described by Tony Jackson[4] followed the major earthquake which occurred in Guatemala in 1976. Some 23,000 people were killed and over a million were left homeless. Relief agencies responded on a big scale, not only with medical aid and help with rebuilding, but also with food. Yet the earthquake had not damaged growing crops, and a record harvest was completed in its aftermath. But food aid to the stricken region still arrived. American field workers cabled their headquarters to stop them sending food, but it still came. The Guatemala government placed a ban on food imports, but ways were found of evading it. The result was that the bottom fell out of the local market for grain, and local farmers sowed less in the next season.

Some people in Guatemala were certainly destitute and needed help in obtaining food. But at least one relief agency met this need by buying food grown locally. This could be done in many other countries where people are hungry, thereby supporting local agriculture rather than undermining it.

Part of the problem here is that the West has adopted a set of beliefs – or world view – in which, as Tony Jackson says, 'the Third World is portrayed as a vast refugee camp, with hungry people lining up for food from the global food-aid soup kitchen. This view is false. Some disasters aside . . . the basic problem is not one of food, but poverty. Free hand-outs of food do not address this problem, they aggravate it.'

That does not mean that we should immediately switch to a tough-minded approach in which food aid is never given. Some countries now depend on it; some can use it intelligently, and there are emergencies where it is essential. But the mode of giving needs radical reassessment so that it complements rather than disrupts local agriculture. And before that can be achieved, western planners and all of us need to be aware of the bias in our conventional beliefs about the world. Part of this is a reluctance to think seriously about poverty and how it might be reduced. Another factor, though, is that the West's own major food problem is over-production, and just as the over-production of weapons is justified by phoney intelligence, so also we maintain a phoney world view about requirements for food. This not only includes the 'soup kitchen' image of the Third World, but also the acceptance of a good deal of organized waste, amongst which the inefficiencies of feeding grain to livestock are the most often quoted.

Food, one might think, ought to be central to the arguments that go on about the earth's physical resources, for although there is over-production at present, this may be rather fragile. In 1973–74, after two successive bad harvests in the northern hemisphere, world wheat prices temporarily doubled and food stocks fell to a level which caused some alarm.[5] Moreover, the resources on which farm output depends are regarded in some quarters as very vulnerable. The West's high grain yields and food surpluses are heavily dependent on fertilizers and pesticides produced by the petrochemical industry. It is sometimes said of western diets that we are 'eating oil'. But much less is said about two resources that are disappearing even faster than oil – agricultural land and trees.

Thus it is salutary to find one lone voice asking, 'should we conserve food like energy?'[6] Logically, 'conservation applies . . . to renewable resources', including land, food and trees, as much as it applies to energy. All these things interlock. Modern meat production is not only wasteful in grain and in energy, but beef is now being produced in Brazil by cutting the rain forest to make grazing land for cattle. This exposes a very unstable soil, and once the trees have gone, it is capable of supporting livestock for only two or three years. The resource costs of Brazil's beef exports are thus very high.

Africa presents an equally worrying spectacle. This is the one major region where, if statistics can be believed, food production has consistently failed to keep up with population growth. During the 1970s, per capita food production fell at an average annual rate of 1.1 per cent.[7] There are some states, notably in West Africa, where farm output may be depressed by the effects of food aid. More widespread, however, is the deterioration of soil and the extension of deserts arising from deforestation.

Food and timber are, of course, both forms of energy, and it is the need for firewood for cooking that is the cause of much of the African deforestation. In many respects, this is the most serious energy crisis of all. Yet in the West we tend to see energy shortages very narrowly in terms of oil. Government research in Britain has looked at farm, food, forestry and paper wastes to see how much motor fuel could be produced if these were all hydrolysed, concluding that 27 per cent of all surface transport requirements could be provided this way.[8] But economic recession has abruptly altered perspectives on such issues. During the 1970s, it was widely predicted that oil consumption could go on rising into the 1990s before a crisis of high prices and scarcity forced the adoption of alternative fuels. Now it is noted that a peak in world oil consumption was passed in 1980, and some commentators think that this may prove to be the all-time maximum.

The recession also offers other reminders that the real energy crisis is not just a matter of oil or even firewood, but of poverty as well. Every winter, some old people in Britain die of hypothermia because they cannot afford adequate heating. And in Tanzania, to take one example among many, there is a hospital that cannot use its kitchens and hot water system, because the equipment, given by German donors, is oil-fired. The effect of recession on Tanzania is that oil can no longer be imported on a sufficient scale. So washing, bathing and laundry in the hospital are done in cold water, and 500 meals a day are cooked outside, on open fires, using wood which has to be brought an ever-increasing distance as trees disappear.

Minerals and energy

So many of the world's problems, it seems, turn out to be problems of poverty, but the habitual preoccupations of many of us are solely with technical matters; that must also be our concern in the next few paragraphs, and the reader with a non-technical interest may prefer to move ahead to page 69. Even adopting this narrower focus, we find strangely different views among the experts. Some say that a major resources crisis is looming, but many economists assert that there is no real problem. Tunnel vision is, of course, a factor. Economists want to get back into the business of managing economic growth, while scientists imagine different kinds of crisis according to what their expertise can cope with. Space technologists envisage an energy regime favourable to building solar power devices on satellites. Nuclear scientists envisage an energy crisis for which nuclear fission, and ultimately fusion, is the only possible answer. Biotechnologists claim that their new techniques are the key. Environmentalists hope that energy shortages will force us to take renewable resources seriously.

The economists can defend their case by referring to the way in which prices affect perceptions of resources. For example, it is sometimes said that aluminium reserves could be exhausted within thirty to fifty years. Yet aluminium is one of the commonest elements in the earth's crust. It is a constituent of clay, and so of most soil. However, extraction from clay would not be economically worthwhile, so this source of the metal is not counted in estimates of reserves. But eventually prices will rise and other methods of extracting the metal will develop. Many ores which are not now counted as part of the reserve will then be exploited.

How far can this argument be extended? Wilfred Beckerman says that for most key metals, the amount available within a one-mile depth of the earth's surface is 'about a million times as great as present known reserves'. Even though we do not have the capability to extract this enormous endowment now, he argues, by the time we need it, 'I am sure we will think up something.'

That last comment tells us a good deal about the fundamental beliefs of those who argue that the earth's physical resources will last indefinitely. They have a remarkable faith in technology, and accuse the scientists who foresee a crisis of scarcity of not understanding society, and in particular, of under-estimating the way human inventiveness can respond to market forces.

Another economic argument[9] is put forward by Julian Simon, who notes that the values and world views which people build into their analyses of resources are illustrated particularly clearly by the way graphs are drawn, and by assumptions about interest rates and prices. Simon presents graphs of his own that represent a longer historical period than is usual, and he plots the price of each major resource relative to average incomes. For both food and energy – that is, for grain, oil and coal – the long term trend up to 1980 is for this price criterion to fall. Yet if there were any constraints on supply, and if shortages were looming, we would expect prices relative to incomes to be rising. Thus again, economic argument seems to show that there is no problem.

Such arguments certainly have a point, and we may agree with the economists that scientists do sometimes misunderstand society. But economists do not always seem to appreciate the laws of nature as scientists formulate them, especially the laws that set limits to what technologists can achieve. An example which illustrates this concerns uranium. As with aluminium, the quantity that exists is far larger than the stated reserves. But it is rather too easily assumed that we can get at it all. There are uranium salts in sea water, for example, but to extract them, one would need a plant capable of dealing with enormous volumes of water, which would itself consume a great deal of energy. When used as a fuel in a nuclear reactor, the uranium obtained would, of course, constitute a new supply of energy – but perhaps less energy than was used in the extraction process. To claim that we can add to the world's energy supplies this way is therefore like saying that we can create energy out of nothing, contrary to the laws of nature.

An alternative world view relating to this is provided by the new discipline of energy analysis. The idea is that apart from doing our accounts in terms of money, as the economists do, we also need to account for energy consumed. Peter Chapman,[10] one of the pioneers of this approach, has used energy accounting to emphasize the difficulty of obtaining uranium from sea water. Others have used energy analysis to study the proposal for a satellite-mounted solar power plant. They show that this would barely succeed in paying back the enormous amount of energy needed to manufacture its components and lift them into orbit.[11]

More conventional, earthbound solar energy devices can potentially do much better (though the performance of some existing ones is certainly poor). They can pay back the energy used in their construction several times over.[12] This contrasts sharply with conventional energy systems, whether for electricity generation or domestic heating, which consume gas, coal or uranium and pay nothing back. It will be evident, then, that energy analysis presents a view of resources which coincides fairly closely with what environmentalists have been saying for a long time: that the only way to develop a sustainable lifestyle is by depending mainly on the sun for our energy.

This idea allows us to get the economists off the hook regarding the extraction of uranium from sea water. Suppose that some way were found of using the sun's energy for this, would we then be able to go on obtaining uranium for reactors indefinitely? One method is to grow plants in sunlight and exploit their ability to extract mineral salts from water in their surroundings. Thus, in 1980, Japanese scientists obtained a 4,000-fold concentration of the uranium from sea water by growing algae.

Other ways of extracting minerals from low-grade ores using biotechnology are already under development. They will certainly alter our perspective on resources, and will ease some of the constraints which resource scarcity could impose. But do they provide the ultimate technical fix for the problems associated with resources, pollution, and limits to growth?

A realistic answer must be grounded in understanding of the accessibility or availability of resources. The real problem is not that any key raw material is likely to run out, but that most resources will become less accessible. For example, clay is a much leaner ore of aluminium than the ores that are currently worked. In that sense, the metal would be much less available if we had to extract it from clay. But if, in the future, the richer ores are insufficient to meet demand, means might be found to exploit some clays economically. But there would still be a need to process large amounts of material, perhaps digging up vast areas of land, so there would be a cost in terms of food production. However, it is rarely made clear how big such costs may be, and it is often argued that damaged landscapes can be restored and noxious wastes disposed of, provided only that sufficient money is spent. Nathan Rosenberg is one economist who believes that ecological pessimism is exaggerated because it underestimates our capacity for 'corrective action by using the tools of science and technology'.[13] He suggests that if there are limits to this corrective action in the United States, this is because too much is spent on armaments, space ventures and nuclear energy, and too little on the environment. That is an economic limit, but in his view there is no technical limit on what could be done to correct environmental damage.

Such argument is not entirely convincing, however, and the clearest account of its defects has been put forward by Nicholas Georgescu-Roegen, a Roumanian mathematician-turned-economist who has lived in the United States since 1948. He comments that most of his colleagues in the economics profession fail to recognize limits to what can be achieved by technology partly because they use a linear mode of analysis, but partly also because they assume that damage due to pollution and mining is always reversible. The latter is consistent with a scientific world view based on classical mechanics; it is inconsistent, however, with more recent thinking based on thermodynamics.[14]

To illustrate this point, Georgescu-Roegen discusses several kinds of technical fix which economists regard as being capable of contributing to a solution of the problem of scarce resources. One is the recycling of metal and other wastes. While this is most emphatically to be encouraged, there are two ways in which it is limited. Firstly, there is an energy cost in transporting scrap metal back to a smelter and melting it down. Secondly, many ways of using materials involve spreading them around, in painting or spraying and in general wear, and this makes them unavailable for recycling. It would seem, then, that for a full analysis of questions relating to resources, we need a way of measuring availability, or rather 'unavailability' – and such is provided by the idea of entropy.

Thus the second law of thermodynamics uses the entropy concept to tell us that as energy is used, it becomes less available. Georgescu-Roegen complements this by proposing a fourth law of thermodynamics which states that as material resources are used, they also become less available; machines cannot run indefinitely because their material frames wear out.[15]

These generalizations clarify a point about solar energy applications. The sun continually replenishes available energy on the earth, but there is no equivalent replenishment of materials. Thus solar energy does not provide a complete answer to the question of resources, and Georgescu-Roegen concludes that there is indeed no complete answer – there is no totally sustainable lifestyle.

The fourth law of thermodynamics also clarifies several issues regarding pollution and environmental degradation. It leads to the observation that although most applications of technology appear to make materials more available and more usable, for example, by irrigating deserts or refining metals, there must always be a cost, not only in money and energy, but also in entropy. The evidence of this cost will be some form of waste, either visible as slag or spoil heaps, or less visible as fumes in the atmosphere or reject heat from an engine. All these things represent resources that have been made less available. And the laws of thermodynamics make it clear that every process in the real world must lead to such effects. A solar energy device may run without any pollution whatsoever, but there is no way of constructing it without producing wastes of some kind, and making some resource less available. All processes, of man or of nature, must add to the global deficit.

This leads directly to the conclusion that there can never be a real fix for pollution. For example, the energy we might use in cleaning up a polluted lake will be dissipated in the atmosphere as a small extra burden of exhaust gases and thermal pollution; and the chemicals we use in the same operation will have been produced at the cost of some extra waste at a distant factory. Solving a local problem merely adds slightly to the global one.

Thus Britain's clean air campaign in the 1950s had a dramatic, almost miraculous effect in cities such as Manchester and London. Pea-soup fogs all but disappeared and the incidence and severity of lung diseases declined. This was achieved by controls on burning coal and by greater use of electricity supplied from coal-fired generating stations outside the cities, with very tall chimneys to disperse fumes. That dealt with the local problem, but we now find that it transferred some of the pollution damage from towns in England to forests and lakes in northern Europe, where sulphur dioxide from British electricity production, carried to the ground as acid ran, has killed fish and is now thought to be destroying many hectares of trees.[16] A better solution now feasible is to burn coal at generating stations on a fluidized bed onto which crushed limestone is also fed; this retains the sulphur content of the coal in the ash. Fixes for pollution are like this; they may contain the pollution (as ash in this case), or they may alter its location, but they cannot prevent the production of waste in some form. The only way to do that is to use less resources, because use of materials of any kind is inescapably linked to the production of waste.

Every nation with aspirations for industrial growth has a problem. To meet the requirement for a clean and pleasant environment for its people, selected lakes and rivers may be cleaned up, and limited areas of good environment protected, but that implies a parallel decision that elsewhere, the environment will be allowed to deteriorate. This may even be planned. One idea is that the United States and other industrial nations should designate national sacrifice areas where mineral extraction, nuclear power plants, heavily polluting industries and waste disposal would be concentrated. Every English county planning officer knows that he must designate some parts of the county for waste disposal, and perhaps for a polluting industry. He can often choose where these areas should be and insist on high standards of management, but he cannot choose not to have them.

Differing world views

Mary Douglas, drawing on her work in anthropology, suggests that all human societies have fears about the environment and pollution. In pre-industrial societies, these fears are rooted in harsh experience of drought and harvest failure, dirt and disease, and they are always expressed in ways which link environmental fact with social values. Where there is illness, that may be because of moral transgression associated with impurity or pollution.

Similar connections are made in the modern world. Some people think about cancer in puritanical terms: they argue that prevention depends on abstinence from smoking, sex and fatty food. An equally puritancal note is struck by the conclusion we have just observed that the only way to prevent pollution is to stop burning fuel and mining metal. But the values of people who worry about environmental problems are concerned with more than puritan morals; they also include a concept of nature in which living things have intrinsic worth and the goal of material development is regarded as, ideally, to find ways of living in harmony with nature. By contrast, the confidence of many economists and engineers in human ability to overcome every problem may often reflect a moral judgement that the proper role of man is mastery over nature.

In other words, although the different beliefs about resources and the environment outlined in the previous section are related to scientific theories about energy and thermodynamics, they are also linked to moral attitudes and values. Thus the three (or four) different world views summarized in table 3 are based on different conceptual frameworks within which the same facts may have different meanings. Although scientists may discover objective facts about the environment, they can never construct an objective view of society's interaction with it, because moral values and the social ethos must always be part of that. Thus Julian Simon ends his analysis of population and resources with a chapter-heading which asks: 'Ultimately – what are your values?'. He points out that there is no scientific truth about whether oil resources are being used up too fast, and similarly 'it is scientifically wrong – outrageously wrong – to say that "science shows" there is overpopulation (or underpopulation)' in any particular region or in the world. To seek authority for value judgments such as these by claiming that they have been established by science is no better than to claim that

TABLE 3   An interpretation of three 'world views' regarding natural resources

Technologists tend to be divided between a majority who take a technical-fix approach and a minority interested in energy analysis or other environmental aspects. The technical-fix approach coincides closely with the views of the 'economists' and is not discussed separately.

	Economics (e.g. Beckerman)[1]	Technology-based		Bio-economics (e.g. Georgescu-Roegen)[3]
		Technical fix	Energy analysis (e.g. Chapman)[2]	
World views as defined by:				
(a) science	classical mechanics	thermo-dynamics[4]	thermo-dynamics[4]	amplified thermo-dynamics[5]
(b) concept of nature	nature as machine	nature as machine	nature as system	nature as process
(c) proper role of man	mastery of nature		acceptance of nature	harmony with nature; restraint
(d) currency of account	money		energy	entropy
(e) time per-spectives	20–50 years, and growth open-ended		20–50 years; equilibrium; limits to growth	100–1000 years? with limits to sustainability
(f) concept of technology	production	construction, innovation	construction, innovation	management of process
Policy implications	nuclear energy, solar satellites, improved mineral extraction		solar energy, biological technologies, modified lifestyle	

[1] Wilfred Beckerman, *In Defence of Economic Growth*, London, Cape, 1974.

[2] Peter Chapman, *Fuel's Paradise*, Harmondsworth, Penguin, 1975.

[3] Nicholas Georgescu-Roegen, *Energy and Economic Myths*, New York, Pergamon Press, 1976.

[4] This is founded on four laws of thermodynamics, as follows: *Zeroth law*, about thermal equilibrium; *First law*, that energy cannot be created or destroyed; *Second law*, about entropy and the availability of energy; *Third law*, about the absolute zero of temperature.

[5] Georgescu-Roegen adds a new law of thermodynamics: *Fourth law*, 'unavailable matter cannot be recycled'.

racial discrimination is scientifically justified by research in genetics.[17] In today's world, ecology is a science which has been abused particularly badly in this way.

Yet it is clear that ecology, like thermodynamics, can provide extremely valuable ways of analysing some of our current problems. Thus while we use the insights of these sciences, we ought to be very clear about why they cannot lead to firm conclusions about future resources that everybody can agree on. Once again, the same facts have different meanings according to the conceptual framework we use; and especially crucial to any such framework is its time dimension. This applies particularly to thermodynamics, where the conclusion about a steady deterioration in the availability of resources seems inescapable, but the rate of deterioration and the time over which problems will arise is undefined. Man's industrial activity has tended to accelerate the process, but we may still believe that the deterioration is so slow as not to matter.

There is also no agreement about what time-scale in the future we ought to be considering. Varying but often vague assumptions about this can make all the difference between optimism and pessimism. Thinking twenty years ahead, many people would agree that there is perhaps no problem about energy resources (except firewood), but that forests, soils and farmland ought to cause us concern. But to think ten thousand years ahead, by which time there may be a new ice age, is for most purposes absurd. On that time-scale, we may even contemplate the possible end of human civilization with philosophical equanimity. Most people want to feel that the future is assured for much more than twenty years, but are not worried for so far ahead as ten thousand. One study group has argued that we have an obligation to future generations to think fifty years ahead, and criticizes most commercial and political planning for its 'horizon blindness' beyond about ten years.[18]

Economists argue that growth is sustainable, but on what time-scale? Environmentalists and energy analysts deny that much further economic expansion is feasible, but argue for a lifestyle that is kept in equilibrium with the environment by attention to recycling and the use of solar energy (table 3). Georgescu-Roegen, though favouring the latter proposals, argues that even this way of living is not sustainable in the long term. Thermodynamics, he reminds us, portrays a universe which must gradually run down, like an unwound clock. Social scientists who have reviewed these issues conclude that, as Stephen Cotgrove puts it, 'views of the future are rooted in systems of meaning that are social constructs and lack any firm objective certainty. They are faiths and doctrines'.[19]

But action and policy need a firmer footing. The lesson can only be that all these beliefs need to be examined with the greatest circumspection. The dominant world view in modern society is that of the economists and the technical-fix brigade. But that does not mean that they have the best evidence on their side – only that they exert most influence over decisions. So Cotgrove warns that this dominant 'faith' has 'no superior claim to truth'. Indeed, simply because it determines most policy within industrial societies, it may be 'in the most urgent need of sceptical reappraisal'. It would seem prudent, then, to take careful account of the arguments of those who talk about ecology and thermodynamics. One sees opposition to their attitudes: 'their view of reality must struggle against considerable odds. But they could be right. And if their vision of the future is closer to reality, and if we ignore it, then the results could well be disastrous . . .'

Georgescu-Roegen carries his analysis of these issues to the point where he formulates specific 'bioeconomic' policy proposals. Recommendations range from population policy to wider use of solar energy. The suggestions are similar to those of other environmentalists except that they are related to thermodynamic concepts concerning the processes of nature. Indeed, they can be taken to imply a concept of technology as the management of process. There are two particular points of significance here.

Firstly, this perspective offers a philosophical basis for the emphasis on maintenance and disease prevention stressed in the previous chapter, for these can both be regarded as means of husbanding the products of artificial or natural processes. Maintenance and prevention are both conservation activities; they both seek to slow the inevitable, cosmic processes of decay.

Secondly, this concept of the management of process calls for some reassessment of the conventional view of technology as being primarily concerned with the engineering of inorganic matter. As one engineer has observed, his profession 'has been going through a phase of rejecting natural materials. Metals are considered more "important" than wood.'[20] The same is true of fuels and chemicals.

But thinking about technology in terms of process involves recognizing that most of the processes which take place in the world are actually biological ones. For example, the plant material produced by photosynthesis each year amounts to ten times as much fuel as the world's population annually consumes. That is not to say that plant material – any more than engineered solar energy devices – can provide all our fuel. Such simple answers are always suspect. But in a broader sense, living plant materials – especially trees – contain many of the processes that need to be managed, and constitute one of our biggest resources. We ought to worry much more about how that resource is being destroyed, whether by cattle ranching in Brazil or by acid rain in Europe and North America. The *Global 2000 Report*, commissioned by President Carter, estimated that the world's stock of growing trees would fall from 327 billion cubic metres of timber in 1978 to 253 billion by the year 2000.

Yet it is a mistake to think of trees and plants as merely material or fuel, for they also contain information in genetic form. It is estimated that something approaching half of the world's genetic heritage is to be found in the tropical rain forests. It is claimed that here are all the resources for a natural biotechnology which, running on solar energy, could provide 'great cornucopias' of substances with 'potential in medicine' as well as timber and food.[21] But trees are being felled and forest lands cleared at a rate of around ten million hectares per year, and the rain forests may be largely destroyed before an adequate scientific assessment has been made of the resources being lost.

Yet even as the importance of biological resources is recognized, we still think in engineering terms about how they should be exploited, with the result that in some places, trees are being destroyed even more rapidly and soil resources are misused with the cultivation of alchohol fuel crops.[22] There is a parallel irony in much solar energy development, where the emphasis is on solar-powered water heaters, solar-electric devices, satellite-mounted power stations and wind turbines. All these depend heavily on conventional engineering and inorganic materials (steel, glass and aluminium), and so on established coal-fired and nuclear-powered industries.

The human potential

One drawback with all the three world views as they have so far been presented is that they do not take account of the end-use of the resources they consider. To complete the picture, we need to ask: what human purposes and goals are served by expanding economies, and what limitations on living are there in more sustainable conditions? How is poverty to be overcome if not by economic growth? It is mainly by answering these questions that we can begin to recognize what choices are available and avoid the impression that some kind of thermodynamic determinism is at work. For even if we accept that the laws of nature set limits on the material aspects of life, there is no absolute, fixed connection between material and social development. As Georgescu-Roegen puts it, what matters most is not the material flow from resources through the economy, but the immaterial flow to human well-being. And one knows that there have been communities living in material poverty where well-being was high. For more than a century and a half, western visitors to Botswana in southern Africa have noted something of this.[23] One described communities there by talking about the 'abiding city' of 'human civilization and enlightenment', and speaking of a 'welfare society', though on the level of 'grinding poverty'.[24]

But romanticized accounts of poor communities abound, and it is only if the part of the description that mentions 'grinding poverty' is tackled realistically that we can begin to make gains in understanding. Grinding poverty usually means insufficient food, ill-health, poor housing, and poor educational opportunity; the conventional wisdom is that only economic growth can raise standards in these areas.

Yet there are places where health and education do seem to advance, even in the absence of much economic development. One such is Sri Lanka. Another is the nearby state of Kerala in south India. Statistics indicate that in both, education at primary level is good (table 4). Life expectancy at birth exceeds 65 years, which is closer to the 73 years that Americans can look forward to than the 55 years more typical of India as a whole. It would appear, then, that high incomes are 'not necessarily a prerequisite for low mortality, a message of obvious significance to other low-income countries'.[25] Equally, it seems that good health and social progress do not always depend on a high consumption of material resources. And this conclusion seems so important that the remainder of this chapter is devoted to a case study, focusing on Kerala.

One of the most sensitive indicators of a community's level of health is the death rate among infants during the first year of life. This reflects the health of both mothers and children, and shows what kind of start in life the new generation is getting. In Kerala, infant mortality fell sharply from the 1950s and into the 1970s. This is illustrated by figure 7, and is an achievement which deserves to be recognized as 'technical progress' just as such as any of the developments plotted as graphs in chapter 2. Indeed, the conventional wisdom is that reductions in infant

TABLE 4 Some indicators of health and economic conditions in Kerala and Sri Lanka compared with the whole of India

	Year	Sri Lanka	Kerala	All-India
Quality of life				
Life expectancy at birth (years)	1981	*67*	*66*	*55*
Infant mortality within one year of birth, estimated per 1,000 live births	1958–9		90	150
	1968–9		66	132
	1978	42[a]	40–50	120–30
Adult literacy rate	1981	70%[a]	69%	36%
Children enrolled at primary school	1977	94%	95%	78%
Demography				
Crude birth rate per 1,000 population	1977	28.5	27.9	32.9
Crude death rate per 1,000 population	1977	7.4	8.3	15.4
Average annual population growth	1971–81	1.8%[a]	1.74%[b]	2.21%
Couples using family planning	1981		29%	22%
Food supplies				
Estimated average energy (calories per person per day)	1972–5	*1950–2250*	*1600–2300*	*1950*
Protein (grams per person per day)	1972–5	*45*	*47*	*55*
Economy				
GNP per capita (equivalent US dollars)	1975	*200*	*135*	*145*
Growth in GNP per capita per year	1960–76	2%	1.3%	
Electricity use per capita (kWh per year)	1972–3		*81*	*97*
Growth in per capita electricity use per year	1972–3		11%	8%

[a] Sri Lanka figures calculated at slightly different dates from those stated
[b] trend markedly to lower growth rate

Sources: United Nations and UNESCO Statistical Yearbooks; also T. P. Dyson, 'Preliminary demography of 1981 census', *Political and Economic Weekly* (Bombay), 16, no. 33 (15 August 1981), pp. 1349–54; R. H. Casson, *India: Population, Economy, Society*, London, Macmillan, 1978; D. R. Gwatkin, in *Food Policy*, 4 (1979), pp. 245–58; P. D. Henderson, *India: the Energy Sector*, Delhi, Oxford University Press, 1975.

mortality in England, also shown on figure 7, were primarily due to such applications of technology as the improvement of water supply and sanitation and the hygienic bottling of milk. One factor that tends to be forgotten is that when infant mortality in England began to fall in the late 1890s, the mothers concerned were the first complete genera-tion to have benefited from universal primary education.[26]

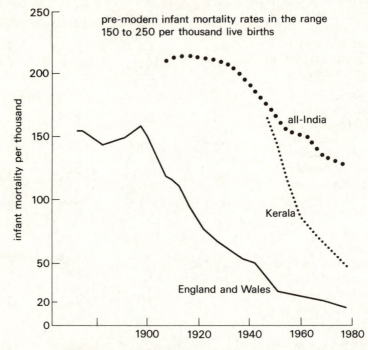

FIGURE 7   *Mortality of infants during the first year of life, in India and in England (Data for India are very approximate)*
Source: Arnold Pacey, 'Hygiene and Literacy', *Waterlines*, 1 (1), July 1982, p. 26.

In both Kerala and Sri Lanka, improvements in water and sanitation have come slowly, and cannot account for the recorded changes in mortality; malaria control, despite its limitations, has had a bigger impact. But primary education and adult literacy have developed rapidly. In 1981, some 69 per cent of the adult population of Kerala were literate compared with an average literacy rate throughout India of 36 per cent. The difference is striking, but how might it affect child health and mortality?

Two factors may be involved, one of an immediately practical nature, and one more intangible. The practical point is that where most mothers are literate, they make better use of the well-staffed network of primary health clinics which is one of Kerala's most notable features. The mother can read her baby's health record when it is filled in at the clinic; she can read family planning posters and the instructions on medicine bottles.

Many experienced people 'view literacy instruction as the best possible means . . . to break the vicious circle of low incomes, high birth rates, and slow development'.[27] But more important than its immediate utility, reading offers a less tangible benefit; it gives the individual ability to control the 'message input' – to stop, think, and reread difficult sentences – and this may allow unused mental abilities to be unlocked. Acquring literacy sometimes makes people more innovative and more self-aware. Sociologists in South America have found a close connection between functional literacy and innovations in the home such as the installation of latrines and medicine cupboards.[28]

In some places where this influence of literacy on levels of awareness and social development has been noticed, it has been spoken of as an awakening or a consciousness-raising effect. In Kerala, it seems, there has been an awakening which has led people to grasp new possibilities in health and social change. One account of a rural latrine-building programme there traced its origins to a women's literacy class.[29] More generally, it is said of Kerala that 'sanitation is good because people are educated'. There may be no modern sewerage in most places, but good standards of public hygiene are maintained because of the care taken over removal and disposal of night soil.

Kerala's achievements in education and health during the 1970s were made against a background of considerable poverty. Most economic indicators were below the average for India (table 4) and Kerala was one of the poorer states in the country. What this may mean to the individual has been vividly described by the economist Leela Gulati,[30] through a detailed record of the lives of five poor women in Kerala. Their employment opportunities are limited, and tasks are segregated by gender. Women may do heavy work carrying bricks, but get paid half the wages of men in the brick-making industry. Work available in the coir industry and in agriculture may be equally badly paid. Yet often women feel that they have no choice but to work if they are to buy food for their children.

Where people are living so near to the limit of subsistence, economic growth as well as redistribution of wealth must be desirable, and fortunately there is evidence of such growth occurring. Rice production has been rising by around 3 per cent annually, and tapioca production much faster. Fisheries do well, and engineering workshops capable of repairing diesel engines for fishing boats have blossomed into small industries. One now manufactures engine components, including valves and crankshafts. Other engineering works have benefited from the rapid expansion of electricity supplies (table 4), helped by investment from the Kerala government, and have gained contracts to make equipment for the Electricity Board. The improvement of health services may have helped also; on the edge of the Cochin-Ernakulam conurbation, a small industrial estate is devoted mainly to enterprises making drugs and medicines. One local entrepreneur began his career as an apprentice to a traditional or ayurvedic physician, and now has modern equipment manufacturing nine different pharmaceutical products.

An industrial survey which reports these instances also mentions a factor which has certainly held back the local economy: almost half the factories covered had suffered through strikes. 'One plastics manufacturer, who had just experienced a 2½ month strike, said that his major aim now was to save for more automatic machinery.'[31] In the rural areas, there has also been a long history of labour unrest. This slowed down the introduction of the new rice varieties associated with the green revolution, and in the early 1970s, deprived Kerala of some of the economic expansion which other states experienced. So bad was the trouble that farmers sometimes left fields uncultivated rather than face the problems of employing labour.[32]

At the root of this unrest were not just the usual campaigns for better wages and for land reform. In addition, from 1964, there was a split in the large local Communist Party. A strongly Marxist faction broke away to form cells in the villages and to promote a continuous mass campaign against employers. Other sectors of society shared the feeling that led to all this, and a parallel campaign within the Catholic Church in Kerala argued that true religion should involve the 'struggle for bread, human dignity and social justice'. Members of the Mahatma Gandhi movement also talked about a 'struggle' aimed at 'organizing the poor to overthrow the bonds . . . of vested interest'.[33] These Catholic and Gandhian groups have complemented the militant labour movement by organizing 'social welfare societies' working mainly with women and seeking to improve women's employment opportunities. They have sponsored literacy classes, handicraft workshops, and an impressive waste-paper recycling project in Cochin-Ernakulam.

All this ferment is reflected by the many flourishing local newspapers representing several points of view – Communist, Muslim and Congress Party. The circulation of papers in the local language exceeds two million, and they are read aloud in village teashops and vigorously discussed.[34] Political turmoil seems thus to have stimulated the growth of literacy. At the same time, a long history of trade unionism, and of political parties representative of the poor, has helped keep governments interested in providing education and health services at the most basic level.

In evaluating Kerala's development, we are faced with some of the same problems as in assessing the Cultural Revolution in China. In both instances, economic growth was badly impeded by mass movements, but political experience and literacy, plus the awareness these brought, may have paved the way for other kinds of development. The parallel with Sri Lanka also needs to be remembered. In one of the most convincing of recent assessments, Davidson Gwatkin notes cultural similarities between Kerala and Sri Lanka which differentiate them from the rest of the Indian subcontinent. He observes that in both states, the 'cultural setting' has fostered 'political systems unusually conducive to popular will. Open elections are prominent features of each society.'[35]

Popular participation is not without its drawbacks, however. Governments in Kerala and Sri Lanka often look ineffective and are frequently overturned by ballot. They are prisoners of popular feeling, and that 'prevents adoption of the tough development measures necessary for growth'. The contrast with the newly industrialized countries of south-east Asia, with average economic growth rates of around 8 per cent during the 1970s – but with governments that are rarely democratic – could hardly be more striking.

Gwatkin suggests that in Kerala and Sri Lanka, the 'tyranny of public opinion' may have led to two specific features of policy: an emphasis on production of food for local consumption rather than cash crops for export, and a welfare system centred on primary health clinics, primary education, and provision of heavily subsidized food on ration. Sri Lankan agriculture has enjoyed considerable success with an average annual 6 per cent increase in rice output during the 1970s – a record which outstrips all other South Asian countries apart from Thailand.[36]

Although food supplies have grown faster than population, statistics suggest a rather low average consumption, especially of protein (table

4). But there is little evidence of malnutrition in either Sri Lanka or Kerala, possibly because diets contain a good variety of foods, and possibly also because, as Gwatkin says, the stress 'on education (especially for women) . . . seems likely to exert influences on diet composition, by helping heighten nutrition awareness'. Above all, there is an equitable distribution of food based on the subsidized ration and fair-price shops.

Government policies in Kerala and Sri Lanka were once 'scorned' by economists of the international community because of the very limited economic growth they produced; attention was focused further east, on fast-growing Taiwan and Korea, Singapore and Hong Kong. But in 1979, Gwatkin reported a change of mood. Policies in Kerala and its neighbour 'are now praised because of the remarkable social achievements for which they are said to be responsible'. Since then, however, there has been a further change. Sri Lanka's policies have moved markedly away from 'basic needs' and towards the encouragement of economic growth. There have been cuts in food subsidies, the establishment of free trade zones, and a massive hydro-electric scheme that has meant the loss of a large proportion of the country's remaining natural forest cover.

The contrast between basic-needs development in south Asia and rapid growth along the East Asian Pacific rim is blurred in another way also. In the latter region, one or two countries have not only achieved economic success, but have also made impressive advances in meeting basic needs, with a considerable reduction in income differentials between rich and poor. Taiwan, in particular, has become known as an instance of 'growth with equity'[37] with its land reforms and rural co-operatives as well as electronics factories and steelworks. Moreover, it is widely observed – in Kerala as well as Taiwan – that with increasing literacy and greater equity, birth rates fall, population growth slows, and there are qualitative improvements in living standards. Mainland China seems to have achieved similar results by yet another route, so again it seems clear that there are no totally unique right answers.

There are no unique formulae for raising living standards, that is, if we consider political systems, or growth rates, or the technical fixes associated with western aid. But one thing common to most countries where equity has increased alongside economic development is that, in one way or another, the interests of low-income groups have been effectively represented in the political process. Sometimes this has come about through peasants' movements, trades unions and organized protest, sometimes through open elections, and sometimes through revolutionary change. Circumstances differ greatly, but nearly always, demands have had to be expressed from below before there is a more equitable allocation of resources – including food, welfare provision, and in agricultural communities, access to land.

In some respects, perhaps, these southern and eastern parts of Asia present us with the most striking image available of the kinds of *choice* which may confront us in the future. Neither the rapid-growth societies of the Pacific rim, nor the basic-needs approach of Kerala may seem ideal. But if we in the West were less preoccupied with the narrowly technical aspects of resources, we might find constructive potential in what these examples demonstrate.

Firstly, we might become more aware that many of the problems of food and energy arise from inequitable distribution, and are not primarily due to material limits. Figures about the current extent of malnutrition throughout the world, or about likely future shortfalls in minerals, food and energy, draw attention to the supply of these commodities, and are often intended to gain support for policies (and technologies) that will increase production. We saw many examples of this in chapter 3. But what matters to people is their consumption, and it is often poverty, inequality, and costly or wasteful technologies which restrict consumption, not fundamental scarcity.

But scarcities there will sometimes be, and a second and more hopeful point of which Kerala should remind us is that a great deal can be achieved by human endeavour even where material resources are limited. As beliefs differ, so many will evaluate these issues differently. But perhaps we should consider that it is our own resources as people that may ultimately matter most.

5

Imperatives and Creative Culture

Imperatives and choice

The prospect of choice with which the last chapter ended implies criteria for choosing. It therefore implies values. In the choices between broad socio-economic options, the values involved are clear enough: wealth on the one hand and welfare on the other. Even the more narrowly technical ideas discussed concerning resources were seen to depend on values, such as the idea that it is good and right for mankind to seek mastery over nature (table 3). But on the more mundane level, 'values which are incorporated in technological products and which guide and inform the actions of technologists and those who direct their work, are either unrecognized, or simply taken for granted'.[1]

Hence the idea has come to be accepted that technology is value-free. People have come to feel that technological development proceeds independently of human purpose. They see it as the working out of a rational pattern based on impersonal logic. Yet in discussing the beliefs which contribute to the culture of expertise and to conventional world views, the three preceding chapters have inevitably implied a good deal about the values that underpin those beliefs. If people use the steadily-rising efficiencies of steam engines as evidence for a linear pattern in technical progress, that is not only because they believe that this provides a realistic insight into the nature of such progress, but also because they value increasing efficiency as something desirable in itself, they value the rationalism and logic which the linear pattern seems to show, and above all, they value technical progress. Yet if people were to assess technical progress in terms of different values, they might present diagrams showing infant mortality more often (figure 7), and graphs representing efficiency in its various forms less (figures 3, 5).

Observations of this kind tend to confirm that essentially materialist, economic values are dominant within the industrial culture, and that among technologists, efficiency[2] and rationality are also important values. But all these are foreground values. They do not engage at all clearly with the impulses and drives in much of the most dynamically creative technology. The latter are impulses often referred to collectively as the 'technological imperative'. But this is a blanket term, and little is ever said to uncover its real meaning and to reveal the background values that support it. The clearest statements about it tend to describe the imperative as the lure of always pushing toward the greatest feat of technical performance or complexity which is currently possible. It then comes to seem an inexplicable, innovative force that cannot be restrained. Discussion of the technological imperative thus reinforces the determinist impression of technical advance that many people entertain. Technical progress, they say, is 'autonomous';[3] the microelectronic revolution is 'irresistible'.[4] And it is made to appear that engineers, scientists and medical men have been taken over by a blind, uncontrollable power which dictates that whatever is feasible must always be tried, and that every new technique found to be practicable must then be applied. 'We cannot stop inventing because we are riding a tiger.'[5]

This way of talking is an evasion. The technological imperative is partly an expression of values, but not exactly the ethics and values openly acknowledged in professional life and presented as a prominent part of the 'expert sphere' of practice in figure 6. The imperative rather refers to more obscure personal 'experience' of technology and to background values. It is important to identify these and try to understand the sense of compulsion to which they lead.

One suggestion is that most talk about imperatives and the inevitability of current patterns of technological advance hides an 'unquestioning acceptance of economic values'.[6] This, however, seems inadequate in that a good deal of high technology, notably in aerospace, goes far beyond anything that can be profitable or can make economic sense. Some people explain that by pointing to the many developments which are politically rather than economically motivated. Supersonic airliners and space exploration are promoted for reasons of prestige. Microelectronics and automation are part of an effort to centralize control over administration and production. For long periods during the 1960s and 1970s, more than half the scientists and engineers working in the United States have been employed on defence projects, aerospace and nuclear power. Such a narrow concentration of politically motivated effort must have its influence.

Yet there are still occasions when technological development seems to escape political control, and when the imperatives behind it go beyond even military requirements as well as economic sense. The biased projections and one-sided world views of the experts sometimes have the effect of manoeuvring politicians into positions they never wished to take. As we saw in chapter 3, that seems to have happened with some aspects of policy on nuclear weapons. The implication is that technological advance is sometimes pursued for departmental reasons relating to the dynamics of particular professions. Sociologists have examined the latter aspect in terms of the way professionals in science and technology seek recognition from colleagues and gain status within the professional community through their achievements in discovery and innovation.

Undoubtedly many factors are at work, and professional interests are important among them. But I believe that if we are to understand why technological development takes such a hold of people that it is discussed in terms of imperatives and obsessions, we need more than what a sociological investigation of such issues can provide. We also need to enquire about the meaning of technology for those people who develop or use it, and that entails enquiry into what Samuel Florman has described as its 'existential' aspect. His proposition is that 'the nature of engineering has been misconceived. Analysis, rationality, materialism and practical creativity do not preclude' personal values or emotional purposes. Indeed, they are 'pathways to . . . emotional fulfilment', and it is the wish to achieve fulfilment that must account very largely for the so-called imperative. 'At the heart of engineering lies existential joy.'[7]

Florman's views are of considerable interest in that he is a practising engineer in New York state, yet his way of explaining the human significance of engineering is less by direct reference to his own experience than through the exposition of literary works that allude to technique and skill. I shall draw on his approach a good deal, but in order to illuminate the relationship between the existential experience of technology which he describes and the economic values which are supposed to be dominant, it is helpful to refer to another author who at one time also aspired to be a civil engineer in New York state, and to that end applied for a post on the Erie Canal. He was Herman Melville, and in his famous novel *Moby Dick* he described how the voyages of a New England whaling ship were promoted by men 'bent on profitable cruises, the profit counted down in dollars'. These individuals felt confidence in employing a certain Ahab as captain to take charge of their ship, because of his single-minded commitment to the whaling life. But what these 'calculating people' failed to see was that Ahab's commitment arose from the 'infixed, unrelenting fangs of some incurable idea' that was decidedly not connected with profit. They did not realize that this idea – the pursuit of one particular white whale of legendary ferocity – endangered their whole enterprise.[8]

So it is with much technology. Industrialists and politicians are often unaware of the infixed ideas motivating the research and development they sponsor. They do not see that the projects on which they employ technical staff may take on purposes for those people which are quite unrelated to the initial economic or other utilitarian goals of the project. These ulterior purposes may not entail the pursuit of a white whale, and usually will not endanger the project, but they may give it a certain momentum and bias that is difficult to resist.

Florman is not the only one to identify these hidden purposes as the pursuit of existential joy. Mary Douglas[9] speaks of the joy that comes through discovering and understanding how systems work. Einstein is often quoted as talking about 'the joyful sense . . . of intellectual power' that can be found through work in science.[10] And it is very clear that in the practice of technology as well as science this goal has its place. There are many technologists who see their work partly as a response to challenge, and seek joy in achievement. Impulses such as this lie behind J. K. Galbraith's talk about 'technological virtuosity'.[11] The phrase 'technological exuberance' expresses the same thing for Herbert York when explaining why scientists in the United States have often pushed the development of nuclear weapons far beyond the requirements of a rational defence policy. In chapter 3 we observed the stratagems by which the weapons experts did this; the pursuit of technological exuberance and what I shall call 'virtuosity values' are part of their motivation. Oppenheimer, for example, is famous for his statement that one invention used in the hydrogen bomb was 'technically so sweet that you could not argue' against its adoption.[12]

It is a great mistake to allow the myth that technology is value-free to blind us to these impulses and values. It is especially mistaken to regard these imperatives as inexplicable and deterministic. Few technologists will acknowledge the 'infixed, unrelenting fangs' that are often the source of their drive; few talk about existential joy. But the fact remains that research, invention and design, like poetry and painting and other creative activities, tend to become compulsive. They take on purposes of their own, separate from economic or military goals. And if technologists feel like this, may one not expect that technical judgements may be influenced? Or that the products of technology may be expressive of such emotions?

The ethos of the 'technology-is-neutral' outlook outlaws all such questions. They are inconsistent with the supposedly detached and objective methods of technology. But such questions must be asked if we are to understand the apparently blind imperatives which appear to characterize modern technology.

Aesthetics and mobility

The one branch of technology in which taboos against talking about values are almost non-existent is architecture. This has close affinities with engineering, but because it is regarded as art, its role in expressing cultural values is openly discussed. And in writing about the products of technology, architects have often noted that 'constructive and artistic genius' shines out with 'as clear a light' from automobiles, aircraft, refrigerators and radio sets as from any building. Maxwell Fry, a leading architect of the 1930s and 1940s, followed these remarks with a comment on the mechanical design of railway locomotives; the earliest ones were clumsy and ugly devices. But as engineers gained experience, more refined types were produced which were so 'purged . . . of clumsiness and inconsequence that they became objects of the most impressive beauty'.[13]

This preoccupation with the relationship of machines and architecture has often been asserted by quoting Le Corbusier's saying, 'a house is a machine for living in'. That has been taken to mean that buildings should be austerely functional, and that architecture should be as emotionally neutral as machines are supposed to be. But Le Corbusier's architecture was not like that; it was vigorously expressive, and lived up to another comment he made, that his aim in building was 'to create poetry'. The initial statement about houses and machines needs another interpretation, therefore, perhaps to the effect that machines themselves are expressive of cultural values; and perhaps also recognizing that engineers, too, can sometimes create poetry, or construct the technically sweet.

Certainly one may see the aesthetic impulse in almost all branches of technology. It has often been clearest, though, in the work of craftsmen. We speak of them as having a 'feeling' for materials and for the products they make; they enjoy the texture of wood and the changing colour and malleability of metals as they are heated, forged and filed. A leading historian of metallurgy has asserted that nearly every early innovation in metal-working technique derived from an aesthetic impulse.[14] Craftsmen improved their skill not only by trying to make a knife or sword sharper, but also in trying to improve its appearance, both as regards shape and decoration.

In more advanced technology, aesthetic judgement is still important, because in the design of machines and structures, not every choice that has to be made is rigidly determined by practical requirements, and not every judgement can be settled by calculation. Thus an engineer responsible for the renewal of sewers in one of Britain's industrial cities, some of them 150 years old, has found himself advocating 'technically sweet solutions, where I have proposed rationalizing the existing disorganized system with new sewers in tunnel'.[15] The plans go beyond what is strictly necessary to deal with the initial problem partly in order to reduce the amount of hazardous work in confined spaces, but partly for other reasons. One is a wish to 'improve the order of things', and do a job such as driving tunnels that seems like 'proper engineering'. There is also a strong feeling for materials, with a clear negative reaction against the plastic panels or *in situ* resin linings that would be used to patch up sewers if new work were not undertaken, and a positive preference for concrete. 'Even better is brickwork . . . Some of the most satisfying jobs recently have been on re-lining of existing brick sewers . . . by a new skin of engineering brickwork (this also was the cheapest solution).'

In view of what has been said about the unattractiveness of maintenance work for many people, this engineer's final comment on the repair of brick sewers is particularly significant. He speaks of a '*strong* satisfaction' experienced in 'restoration and adaptation of the existing structure . . . In fact, planning and executing a good repair on a trusted existing structure (or machine) which has already seen worthy service is to me more satisfying than creating from scratch.'

This feel for materials and the visual satisfaction of knowing that 'if the job looks right it *is* right' may be shared by many engineers but only occasionally, with some of the most famous ones, is it recognized and discussed. Thus Robert Maillart's brilliant designs for concrete bridges have been studied with a view to identifying the aesthetic judgements involved and the connection between technical logic and the artistic features. The term 'artist-engineer',[16] borrowed from a Renaissance context, has been applied not only to Maillart, but also to F. W. Lanchester, designer of automobiles, and to I. K. Brunel, builder of ships and railroads.

Everybody who uses the products of technology may share in enjoyment of their aesthetic qualities, just as everybody entering a building may share something of the architect's feeling for its structure, space and decoration; and people's everyday enthusiasms are reflected by their purchase and use of equipment. I can speak best about personal experience: equipment which I find enjoyable and even stimulating to use includes a calculator, slide-rule, bicycle, camera, typewriter and xerox copier. But the enjoyment and sometimes exhilaration I feel stems not only from the good aesthetic design of most of these products, but also from the way they enlarge my personal capabilities. The bicycle quadruples the speed I can travel under the power of my own muscles; the xerox copier, hiding corrections, turns my inadequate draughtsmanship into presentable illustration.

But beyond these experiences of aesthetic pleasure and enlarged capability which people derive from technology, another source of enjoyment is associated with having mechanical power under one's control, and of being master of an elemental force. The teenage enthusaism for motorcycles reflects this. Many farmers, it is said, buy larger tractors than they really need, to the detriment of soil structures, because of the pleasure they get from using such powerful machines. Some automobiles are designed to appeal to this impulse; others, of more modest power and pretension, seem designed mainly to enlarge personal capability, like the machines mentioned in the previous paragraph. The dominance of the automobile in the western way of life is not due to blind imperatives, but to the fact that its usefulness is complemented by these two very considerable satisfactions. Exhilaration in speed and power, and the desire for mobility, have perhaps always been part of the human personality; and as Florman says, 'Technologists, knowing of this desire, were, in a sense "commissioned" to invent the automobile. Today it is clear that people enjoy the freedom of movement of which they had previously dreamed.'[17] In most invention, basic human impulses like this precede the technological development. Dennis Gabor talks about 'archetypal human desires' which include the wish to communicate at a distance, to travel fast, to fly.[18]

The impulse to fly, especially, was evident long before aircraft were possible. Perhaps from the time of the Icarus legend, and certainly since Leonardo's sketches of flying machines and Kepler's discussion, in 1610, on flying to the moon, people were attracted by the prospect. Just before 1800, though, this vague impulse became a practical intention, with early hot-air balloons, and with George Cayley's work on fixed-wing gliders pursued with great persistence from the 1790s till 1852.

Economic aims and the profit motive seem quite irrelevant to all this; the imperative here is clearly rooted in non-economic 'virtuosity values', even sporting impulses. Practitioners of hang-gliding today view Cayley as partaking in their sport, and say the same of Otto Lilienthal, another early flyer who experimented with a form of hang-glider during the early 1890s. Such equipment could not conceivably have had a utilitarian purpose: its development, one may argue, was a cultural enterprise, related to the primitive impulse to fly, and to craft skill and aesthetics:

> To learn wind's whims by touch, Lilienthal
> became bird –
> willow wands, waxed cloth, stretched
> resonant as a drum
>
> > leapt, sprinting
> > wings spread into
> > wind's lift
>
> . . . Craftsman, fashioning
> shapes of air . . .[19]

Orville and Wilbur Wright began their experiments on flying after reading about Lilienthal's efforts, and at first they worked solely with gliders. But one writer on gliding and man-powered flight suggests that the Wrights' goal altered fundamentally at the point where they first mounted an engine in one of their gliders. That gave the machine potential to become a thing of economic utility; and this change, it is alleged, diverted technological development away from the sport of 'pure flight', delaying its progress.[20]

The bicycle, the automobile and the aeroplane developed at about the same time using related technologies. In the United States there was a boom in bicycles in the 1890s; it resulted in a string of small workshops including the cycle repair shop established by the Wright brothers in 1892. Techniques for making light-weight frames and for precision manufacture of gears and bearings arising in these workshops were later useful in automobiles and aircraft as well as bicycles. The interactions within this family of industries and the many innovations produced, had very much the character of a movement in technology such as those discussed in chapter 2. There were not only technical links between the three new modes of transport, but they all had similar purposes also, concerned with the mobility of the individual. One expression of this was the interest in using the new vehicles in sport and for establishing speed or distance records. In every respect, there was the sharpest possible contrast with the economic purposes of the railways and steamships produced by an earlier phase of innovation in transport.

The discovery of new dimensions of mobility which generated such enthusiasms for bicycles and aircraft, was experienced in a new form when snowmobiles of the type described in chapter 1 began to appear in the 1960s. One user speaks of an 'almost animal sense of freedom when you realise that the thing can go practically anywhere – shooting up snowbanks . . . across frozen lakes'. There is an immediate contact with wind and weather, a sense of 'anarchistic mobility', and also, as with bicycles and gliders, 'the machine becomes an extension of your body and your senses'.[21]

With all these innovations, then, talk about 'technological imperatives' may disguise a whole range of other impulses concerned with aesthetics, materials and mobility. Especially significant is the impulse to master and manipulate elemental force. One sees this with automobiles and motorcycles, but nowhere has it been more evident than with the steam engine. Over the three centuries of its development, this machine has been celebrated in many ways. Steam locomotives and steamships appear in Turner's paintings; factory-owners built engine-houses like temples or chapels in which to install these highly prized machines; and engines were given names, of which two of the earliest recorded, *Resolution* and *Adventure*, speak volumes about impulses in technology. On the railroads, 'Steam locomotives were a sublime demonstration of man's partnership with the creation. The skill of designers . . . the courage and confidence and sure-touch of drivers;

the rippling muscles and the sweat of firemen; the combination of fire and water and coal and sparks and speed – it was drama and spectacle and poetry rolled into one . . . the embodiment of energy and power.'[22]

But locomotives also take the idea of mastery of nature into another dimension. Walt Whitman not only addressed one as 'Fierce-throated beauty', and 'emblem of motion and power', but called it 'pulse of the continent',[23] referring to the role that railroads were playing in his time, opening up the American West for settlement and helping push back the 'frontier'. Railroads are not often seen in that light today, but frontier values are still an important part of thinking about technology, especially in space and in the few remaining unsettled, unexploited regions of the earth. In America's arctic north, our culture is one in which man is encouraged, 'to conquer the frozen and waste spaces that he sees, with roads, mines, drilling rigs, gas wells and pipelines. He dreams of the technological conquest of the northern frontier'. In the vast, sparsely inhabited interior of Brazil, another of the last frontiers is to be found,[24] and trees are destroyed and roads are built so that it too may be conquered.

Cathedrals of power

The existential joys of technology would seem from all this to extend from the quiet, aesthetic satisfactions of craftsmen to the exultation of the driver of a speeding steam locomotive and the adventuring spirit of frontier conquest. These enjoyments are sanctioned and celebrated by aesthetic ideals and other 'virtuosity values' that claim intrinsic merit for technological endeavour, independently of its utility or economic benefits. Among these values is the idea that it is right and good for man to seek mastery over nature, and that this can be a goal in its own right.

Some commentators see this 'project of conquering nature' as taking shape during the scientific revolution of the seventeenth century, inspired by the voyages of discovery made from the time of Columbus onward, inspired by humanist views of man as separate from and superior to nature,[25] and using ideas also from the Biblical creation myth, in which Adam and Eve were told to 'subdue' the earth and 'have dominion . . . over every living thing that moveth upon the earth'. All three strands of thinking were important to Francis Bacon, when, around 1620, he wrote about an idealized scientific community whose name, 'the College of the Six Days' Works', deliberately recalled the Bible story. Its objects were the extension of power over Nature, and 'the enlarging of the bounds of Human Empire, to the effecting of all things possible'.[26] There we have a statement of frontier values and the technological imperative all rolled into one.

Bacon has been much quoted by technologists with an interest in expounding the cultural significance of their work, though they often forget that he wrote about compassion and discipline in the use of knowledge as well as mastery of nature. Another approach one finds in discussions of the non-economic significance of technology is the identification of its products with archetypes whose cultural achievement is widely recognized. Visiting Liverpool docks at a time when they were the largest in the world and still being extended, Herman Melville said that their 'extent and solidity . . . seemed equal to what I had read of the old Pyramids of Egypt'.[27] In 1913, Walter Gropius wrote about American grain silos which 'can stand comparison with the constructions of ancient Egypt'. Such parallels usually indicate admiration, but Lewis Mumford has said that the objective of the pyramid builders 'was as irrational as our own phrenetic dedication to nuclear weapons and spacemanship'.[28]

Although Melville did not write primarily about technology, he was very aware of the significance of its archetypal achievements, using them as symbols more comprehensively than almost any other author, and linking them with images of fire and with the Prometheus legend. Thus he referred not only to pyramids, but Roman aqueducts, medieval cathedrals and the steam engine. All these symbols are now common currency, and the cathedrals have been quoted especially often in discussions of the aspirations of technology. One reviewer asked of London's Battersea Power Station in 1934, 'Is it a cathedral?' Its architect, Giles Gilbert Scott, was better known as the designer of Liverpool Anglican Cathedral, and was later said to have built two cathedrals, 'one for God, one for Electricity'.[29] Similarly, in 1954, the Capenhurst nuclear fuel plant was described as a 'temple to whatever muse it is that gives inspiration to engineers'.[30] Such comments became even more frequent after an article on large-scale science in the United States was published by Alvin Weinberg in 1961. This pointed to 'the huge rockets, the high-energy accelerators, the high-flux research reactors' then being built as, 'symbols of our time', just as Notre Dame is a 'symbol of the Middle Ages'.[31]

In 1969, Robert Jungk took up the same theme in writing about the CERN high-energy particle accelerator located near Geneva. Echoing Weinberg, he said that this is 'one of the great cultural achievements of our time, the contemporary counterpart of the temples of antiquity, the cathedrals of the Middle Ages'.[32] Three years later, Peter Medawar was talking about how one could claim that a space probe, 'like a cathedral . . . is economically pointless, a shocking waste of public money; but like a cathedral, it is also a symbol of aspiration towards higher things'.[33]

This may offer further clues about imperatives in technology that go beyond utilitarian goals. Any account of, say, a particle accelerator, a nuclear power station, or of the Eiffel Tower has to face the existence of such goals. The latter, in particular, is famous, 'not for its usefulness but its symbolism'. It was the outcome of a hankering widely felt in the nineteenth century to build a tower reaching a symbolic 1,000 feet (or 300 metres) above the ground. Such a construction had been considered for the Philadelphia centennial exhibition in 1876. But it was Gustav Eiffel, a French railroad engineer, who achieved it. Mentioning pyramids and cathedrals, one writer on the Eiffel Tower suggests that all these monuments reflect 'a compulsion constant throughout history to thrust mighty structures toward the sky in moments of special pride.'[34]

Many people, however, may be inclined to dismiss these references to cathedrals, and the snatches of poetry quoted earlier, as emotional froth that has no relevance to the real, practical rationality of technology. It may be, though, that we ought to recognize that the culture of technology comprehends at least two overlapping sets of values, the one based on rational, materialistic and economic goals, and the other concerned with the adventure of exploiting the frontiers of capability and pursuing virtuosity for its own sake.

These two sets of values can co-exist so long as they do not set up conflicting demands. The adventuring, virtuosity-seeking spirit is felt to be admirable so long as it can be paid for, and nobody's security is threatened by it. People have been happy to watch television coverage of landings on the moon and space-shuttle test flights. But when it comes to nuclear energy or weapons, the risks are less remote, and enjoyment of adventure in mastering nature seems less appropriate. Yet nuclear energy must be attractive to many engineers and scientists as perhaps the ultimate example of human mastery over elemental force. Even the test explosion of a nuclear bomb, brighter than a thousand suns but still under the scientists' control, is something to glory in as 'superb physics'.[35]

It is significant, however, that arguments about civil nuclear energy are never couched in these terms, but always refer to economics and utility. Benefits claimed for nuclear power are cheap electricity and a way of meeting anticipated shortages of energy. Risks regarding safety and health are also put into an economic context – they are said to be much less serious than many risks posed by chemical industries, or by air travel, and are sometimes assessed in terms of life insurance.

One result is that the arguments deployed in both sides in the controversy always seem a little forced – they carry too heavy a load of unstated value judgements; they are proxy arguments that stand in for more significant issues which nobody wants to acknowledge. Some of these issues are political, but some are associated with adventuring imperatives that belong to the cultural dimension of technology. Questions of whether or not electricity will be marginally cheaper are mostly irrelevant. Those who argue against nuclear electricity generation perhaps ought to be saying that the power plants are as futile and unnecessary as the Eiffel Tower. Those who favour it probably have a conviction that sometimes Eiffel Towers are crucially important to the future of man's technological enterprise.

One distortion arises, though, from the double-edged cultural significance of the Eiffel Tower, as of the Apollo moonshots. These are both very great human achievements; each can be regarded as 'a small step for mankind'. But both projects were also expressions of national pride, capable of being exploited for political purposes. With nuclear technology, this second aspect seems to have become very important. The over-riding 'motivation' in nuclear development has nearly always been 'pride', according to one recent review, especially pride of the sort called patriotism. So in France, which in the 1980s has the world's most successful civil nuclear power programme, the public is 'as suspicious of nuclear power as the next nation', but '. . . sees its success as one of the great symbols of French strength'.[36]

Critical scientists have argued that the French programme consumes almost as much energy as it produces; they have alleged that the uranium enrichment plant at Tricastin on the River Rhône uses all the electricity output from four nearby nuclear reactors; and even an official report prepared under the supervision of J. M. Bloch-Lainé 'raised severe doubts' concerning the latest series of orders for nuclear plant. Other experts have said that the nuclear reactors will be white elephants by 1990, immobilized by uranium shortage and lack of demand for electricity.[37] But all that is to judge the programme by purely utilitarian criteria and to forget that it also has a political and cultural purpose: it was initiated for 'La France', not only to produce electricity, and just as an American motorist may in the 1960s have bought a much larger automobile than he really needed or a farmer may enjoy using a more powerful tractor than he requires, because of the sense of mastery over the elements it gives him, so a nation may feel a cultural imperative to support a larger nuclear power programme than is economically justified.

In Britain, the 'cult of sheer size' and of technological complexity led, in the 1930s, to 'uneconomically over-size ships like the giant Cunarders', and after 1945, to 'grotesquely uneconomical planes, of which Concorde is simply the last of a long line.'[38] In architecture, there have been heavy commitments to 'what is expensive, but technologically "sweet" ', such as the high-rise flats built in many inner cities during the 1960s; and in medicine, heart transplant operations seem to reflect similar values.[39]

Much the same has been said also about Britain's nuclear energy programme; in 1965, five new power stations were ordered based on a relatively untried concept, the advanced gas-cooled reactor (AGR). There were many difficulties and delays during construction and costs mounted. One economist who carried out a detailed study was moved to say that the decision to build the five AGR plants could rank with the Concorde airliner programme in the exent of the financial loss that resulted. These were the 'two worst investment decisions ever made'.[40]

It is easy to be cynical about such projects and to feel that when references are made to cathedral archetypes, these are merely rhetorical excuses for financial irresponsibility. But for many people, Concorde is a most beautiful and exhilarating aircraft, and the AGR a creative step forward in reactor design, especially regarding safety. Judged by values that have to do with technological virtuosity, these things may seem entirely admirable. And considered also as part of the 'project of conquering nature', the cathedral analogy is entirely appropriate. With upward-thrusting, gravity-defying lines and elaborate buttressing against wind pressures, a cathedral structure may seem to conquer elemental force just as surely as Concorde; it is also, very often, a marvellous feat solely in terms of artistic creativity. So, it may be claimed, is much of today's high technology.

Cynicism, symbolism and creativity

David Dickson has observed that although we may sometimes hear discussion of, 'technology and art', we rarely hear anybody talking about 'technology as art'.[41] Yet it is not only the apparent usefulness of technology that impels us to develop it. There are imperatives that drive us beyond usefulness, though as we have seen, efforts to explain them get us into difficult areas. Aesthetic satisfactions may be easy to understand, but when people talk about the cathedral-building impulse as 'aspiration to higher things' one may suspect an evasion. What higher things? Are we merely seeking an 'endless process of further technological triumphs'? If the ultimate goal is human happiness, may it not be that this 'lies in the quest' rather than in achievement? 'And is not this sort of quest itself a kind of enchantment?'[42]

The medieval cathedrals themselves point toward two kinds of interpretation. On the one hand, they may be simply accepted as artistic achievement, representing the New Jerusalem to the people of their day, built to the glory of God, and reaching up to heaven. On the other hand, though, they may seem to have a mainly political significance, connected with inequalities and power relationships in the society that built them.

Both kinds of interpretation have been applied to technology and the sciences. There are some scientists, according to Einstein, whose interests are contemplative, and who discover in their work 'the silence of high mountains . . . built for eternity'. Quoting this, Robert Pirsig asserts that the same is true in technology. 'The Buddha, the Godhead, resides quite as comfortably in the circuits of a digital computer of the gears of a cycle transmission as he does at the top of a mountain or in the petals of a flower.'[43] And Samuel Florman, the civil engineer, refers to William Golding's novel *The Spire*, which explores the motivations of cathedral building, and adds the comment that: 'Not only cathedrals, but every great engineering work is an expression of . . . purpose which cannot be divorced from religious implications . . . every manmade structure, no matter how mundane, has a little bit of a cathedral in it, since man cannot help but transcend himself as soon as he begins to design and construct . . .'[44]

It may be that part of the reality of the technologist's experience is truly a sense of reaching out toward the transcendant, but for those intent on political explanations, this is at best self-deception. More usually, indeed, the high-flown language will be regarded as little more than the symbolism of advertising. In the 1930s, electricity supply engineers liked their generating stations to be 'cathedrals of power' because they wanted 'symbols of the prestige and modernity of electricity'.[45] Today, the symbolism used in publicising a new automobile, with its play on virility and status, is only too familiar.

In discussing these matters, David Dickson notes that the building of the medieval cathedrals served to reinforce the hegemony of the church authorities. The cathedrals were a means of political control over vigorously developing urban communities, providing a carefully regulated outlet for their wealth, enthusiasm and civic pride. Similarly, a successful space exploration or nuclear project can today give a sense of pride and purpose to an individual nation, and can provide some distraction from more divisive issues.

Civil engineering in Nazi Germany presents a specific example. Giant aircraft hangers made 'the most daring and successful use of new material like shell-concrete'. Autobahns were built 'with magnificence and sweep which can leave no one unmoved. Germans, seeing these things and being proud of them', felt that the regime which produced them 'must be worth something'.[46] Dickson notes that heavy industry had 'an almost mystical significance' in the early years of the Russian revolution, and comments that the '*significance* attached to technology' under these circumstances often 'disguises the exploitative and alienating role technology plays' within industrial societies. That seems especially relevant to some of the euphoric language currently being used in Britain about information technology (IT) and microelectronics; 'there's no future without IT', one booklet asserts, in a propaganda exercise initiated on instructions from the prime minister, Margaret Thatcher.[47] One is reminded of another British prime minister, Harold Wilson, who aimed to promote the 'white heat of technological revolution'. That was the prelude, in 1964, to ambitious attempts to reform government administration of technology, and to promote a range of specific projects, including aluminium smelters linked to nuclear power stations. But the portrayal of technology in terms of virtuosity – the mastery of an elemental 'white heat' – seemed in the end symptomatic of a lack of economic realism.

Such experiences have led to a good deal of cynicism about grandiose claims concerning high technology. One engineer, suggesting a classification of technological achievements, ranks supersonic aircraft, 'very large' power stations and the space race as 'cosmetic machines' and status symbols.[48] Satirizing the same types of technology as they might emerge in a solar age, the Belgian group, Mass Moving, exhibited a particularly remarkable 'cosmetic' machine at Bath in 1974. All shiny metal and complex pipework, it had a large parabolic reflector focusing the sun's rays onto an elaborate boiler. But the creative pretensions of the equipment were out of joint with its practical utility. When it was set to work, it laboured mightily to raise a head of steam – but then used it only to blow a tin trumpet.

Exposure of the false pretensions and phoney symbolism surrounding high technology may often be wholly justified. But there is a third interpretation of the claims that are made and the archetypes evoked in such discussions which deserves attention. This is that they refer neither to any transcendant goal, nor to symbols disguising political aims, but that they simply celebrate the human drive and creativity behind successful innovation. When J. K. Galbraith uses the term 'technological virtuosity', he is chiefly referring to the way technical creativity may be pursued as a goal in its own right. He notes that while the primary goal for any industrial corporation must be economic expansion, technological virtuosity is an important subsidiary goal. This again may be partly for political reasons, because virtuosity attracts customers and helps secure the interest and loyalty of staff. But innovation which demonstrates virtuosity has 'standing in its own right'. As in the scientific work of a university, prestige accrues to the visibly creative organization. And while, in the American corporations Galbraith had in mind, this goal may be held in balance with economic aims, in parts of British industry its over-emphasis may have been a crucial flaw. Chemists working for the chemical firm ICI have remarked: 'We think of ourselves as being a university with a purpose.' Not surprisingly, ICI tends to compare badly with Du Pont in financial and marketing skills. Excellent aero engines have been developed by Rolls Royce, but with some serious financial hiccups, because 'ever since the First World War, commercial values had not been allowed to intrude upon dedication to technical perfection'.[49] Investigating the slow development of electricity supplies in Britain during the 1920s, an official committee found that, 'many of the industry's senior men still accord pride of place to engineering, with the emphasis on hardware and its technical efficiency, rather than on financial and sales questions and service to the consumer'.[50]

Several engineers with whom I have discussed this question speak of I. K. Brunel as the great exemplar of their profession. His career presents many instances of the pursuit of technical ideals regardless of more mundane considerations. Clifton Suspension Bridge, he said, should above all things be 'grand'. The magnificent Great Western Railway, he insisted, must have the broad-gauge tracks which he saw as ideal, despite commercial objections. The *Great Eastern* steamship, built between 1854 and 1860, was to be the largest and most luxurious ship ever constructed; it displaced 32,000 tons, and certainly was the biggest ship built until the very end of the century.

Near the end of his life, during two visits to Rome, Brunel spent many hours on his own in St Peter's. He was not a religious man in any conventional sense, and these solitary contemplations provoked comment from his travelling companions. But his biographer points out that a great deal of the spirit in which Brunel himself had worked was expressed there;[51] under the dome designed by Michelangelo, Brunel probably felt a strong resonance with his own goals and values.

To suggest that Brunel saw St Peter's as symbolic of the impulse that had gone into his own work is to say no more than Alvin Weinberg admitted in comparing nuclear accelerators and reactors to the cathedral of Notre Dame; it is to say no more than Samuel Florman, the civil engineer, who sees every great structure as a 'diagram for prayer'.

Yet it is precisely here that a major dilemma lies. For while we admire Brunel's steamships and bridges, and while we applaud Apollo rockets and jet aircraft, we are guiltily aware of the wasted resources and environmental damage for which many such projects are responsible. We are aware that some of the funds used could be more directly used to relieve suffering and give benefit to people. But so highly do we tend to value creativity that it is not in our nature to place limits on it. We tend to feel that the innovative impulse should never be burdened with too much political restriction or economic parsimony. The word creativity always evokes approval, never distrust; the need for it to be balanced by responsibility is not often stressed, and we do not seem to notice that it is a thin line indeed that separates the wholly admirable artistic or innovative impulse from the arrogance of an individual on his personal ego trip.

But if we go to the other extreme, and denigrate engineers for the apparent irresponsibility of some of their projects, we are failing to recognize the conflicts that they themselves feel between unbridled enjoyment of their creativity and a wish to benefit mankind. That Brunel felt more than just his awesome creative drive is shown by many small incidents, but most notably by his reaction to the dreadful conditions faced by the wounded of the Crimean War. He designed a prefabricated hospital of a thousand beds, and pushed its construction forward so vigorously that it was manufactured, shipped from Britain, erected and equipped within nine months.

In 1974, a notoriously pessimistic report on Britain's industrial future recognized something of the significance of all this. Too many prestige projects, and a lack of economic realism in engineering, had damaged the nation's prosperity. There was also the anxiety shared by the whole of the West about energy supplies and the environmental consequences of the economic growth we all wish to see. The dilemma that had to be faced was one which the writers of the report saw expressed in the virtuosity values of Brunel. In this 'near mythic "Great Engineer" ', they suggested, 'we see the Promethean figure' who gives us cause for admiration but also for 'persistent anxiety about where our civilization will lead us'.[52]

6
Women and Wider Values

Contrasting sets of values

In the ancient world, the achievements of those who today we would
call technologists were sometimes celebrated in legends that described
marvellous feats by mythical metal-smiths and the drama of their
flaming furnaces. The Greeks' artisan-god, Hephaistos (known to the
Romans as Vulcan), was often portrayed like this in stories that date
from the bronze age. When he made a great shield for Achilles, he had
twenty bellows working:

> and twenty Forges catch at once the Fires;
> . . . In hissing Flames huge silver Bars are roll'd
> And stubborn Brass, and Tin and solid Gold.[1]

Hephaistos was widely renowned for his 'craftsmanship and cunn-
ing', and legend had it that technology among humankind began when
Prometheus stole fire from Hephaistos and gave it to man. But one
other deity the Greeks linked with technical skill was the goddess
Pallas Athene (or, to the Romans, Minerva). She stands for the intellectual and moral qualities required in practical work, and for meticulous craft skill. Homer wrote of a carpenter who was 'well versed in all
his craft's subtlety' through Athene's inspiration, and mentioned a
goldsmith who was taught his trade jointly by Athene and Hephaistos.
In other passages, Homer wrote about the practical skills of real women
such as the aristocratic Penelope, known for her weaving and her
wisdom, and others who collected and prepared medicinal herbs.

There was real admiration for the skills of women, though it was
always made clear that these were of a different order from those of
men. On one of the Greek islands there was a community where the

men's 'extraordinary skill in handling ships at sea is rivalled by the dexterity of their women-folk at the loom, so expert has Athene made them in the finer crafts'.

Denied the right 'to a heroic way of life, to feats of prowess, competitive games, and leadership of organized activity of any kind',[2] women excelled in painstaking craft work and socially useful skills – spinning yarn and grinding corn in the handmill. While Homer shows great appreciation for what they did, the contrast with the spectacular achievements of men using furnances and forges, weapons and ships, is very striking. Here, perhaps, we may see two parallel sets of values concerning practical skill and 'technology': one seems rather like the set of values concerning adventure and virtuosity we examined in the previous chapter, while the other is more closely attuned to basic needs and human welfare.

Parallels have been drawn between Homer's world and the life portrayed in roughly contemporary parts of the Bible. The book of Proverbs ends with a poem in praise of the good woman, who rises early, 'while it is yet night' to prepare food for her family. 'She layeth her hands to the spindle, and her hands hold the distaff . . . She openeth her mouth with wisdom; and in her tongue is the law of kindness . . . Beauty is vain but . . . let her own works praise her.'

However, while the work done by women was widely appreciated, other types of work were despised. In Homer's world this included trade and the routine manual work done by men. This attitude was shown by the ambivalent way Hephaistos was portrayed. Surrounded by the flames of his furnaces, his virtuosity and skill are heroic. In other contexts, though, he is a clumsy, inarticulate working man. Similarly, while Penelope's husband Odysseus captained his ship on a voyage of exploration and adventure, he was a heroic figure. But when somebody mistook him for the skipper of a merchant ship, that was an insult. In other words, distinctions were actually made between three kinds of practical skill, carried out by three kinds of people – women; merchants and working men; and adventurers, armourers and warriors.

Homer described aspects of the human character that go very deep, and three thousand years later, we may wonder whether similar attitudes are not still widespread. It is certainly significant that feminist writers, concerned with women's work roles in the modern world, have taken considerable interest in historical and anthropological evidence from traditional societies. Susan Walker, an archaeologist, points out that with regard to ancient Greece, we have 'few references to the way in which women spent their lives, [but] more accounts of how men wished their lives to be spent'. That must be taken as warning that the comments made in previous paragraphs may reflect men's values and aristocratic men at that. What Homer does not describe is how most women 'worked the fields as they have done throughout history',[3] and for an appreciation of what is involved in this, we can most usefully turn to studies of the role of women in modern developing countries.

A statement often quoted is that, in much of Africa, women are responsible for growing three-quarters of the food that is eaten in rural homes. Some such estimate is relevant to much of Zaire, for example,[4] and the figure quoted for Tanzania is that women grow between 60 and 80 per cent of food. Reviewing conditions in the 1960s, Ester Boserup found that in sub-Sahara Africa, more women than men were undertaking agricultural work, and the women were usually working in the fields for longer hours than the men.[5] She also stressed that farming in Africa is not a family enterprise: men and women work independently of one another, growing different crops. Women usually grow the basic food crops for family consumption, while men who are farmers grow crops for sale.

For example, Margaret Haswell[6] has described a village in the small West African state of Gambia where, in the 1940s, the staple food crop, rice, was grown on natural swamp land by the women, while the men cultivated higher fields to grow groundnuts for market. The women had to work very much harder than the men, but received little help from them. During the growing season, women would have to fit in extra work in the fields by cutting down on the time spent on other tasks such as preparing meals and fetching firewood and water. Thus families were less well fed during the growing season than at other times.

Margaret Haswell followed up her study with subsequent visits and found that by 1974, men were tending to concentrate even more on cash crops (mainly groundnuts), and some were leaving the land for paid work elsewhere. Thus women were even more on their own as producers of food, and where there was ill-health, families were short of labour and were going into debt to buy food.

This is the human context of the declining per capita food output in much of Africa that was noted in a previous chapter. The irony is that where agricultural development is planned by governments, this is usually with the aim of directing more farm produce into the market economy. Such policies result in a degree of economic growth, but achieve this by encouraging male farmers with their cash crops –

groundnuts, coffee, cotton – while usually offering no support at all to the female farmers who grow their families' food. The work of these women is dismissed as 'subsistence activity' or 'gardening'. Even in some African countries where most farmers are women, agricultural policies may be based on the assumption that all farming is done by men. One consequence is that in communities such as the Gambian village mentioned, nutritional standards fall as economic growth proceeds. What makes sense in terms of economics does not always make sense in the context of family welfare.

This represents more than the tunnel vision of specialist economic planners. It also demonstrates a conflict of values between economically-oriented development and nutrition-oriented agriculture.[7] One might even argue that agriculture is an instance of technology which takes different forms according to the values built into it.

It is of further significance that, when an agricultural task is mechanized, responsibility for it will often pass from women to men. For example, the milling of grain by hand was not only a women's job in Homer's time, but also in 95 per cent of all communities throughout the world where the technique has been recorded.[8] But where power-driven mills are introduced, the work is usually taken over by men.

In Lesotho, where numerous men are regularly absent from home, working in the South African mines, women may drive tractors and do the ploughing, but elsewhere in Africa, this is exceedingly unusual. In Asia, women are less often farmers as such, and work in the fields mainly as labourers, but are equally affected by mechanization. In Java, harvesting has 'switched from being a female task', and where rotary weeders have been introduced, or rice hulling machines, large numbers of women have been displaced by 'smaller gangs of hired males'. Very often, then, women may simply be left with tasks not affected by technological innovation.[9]

In some circumstances, improvements in hand-tools or water supplies can be identified which do not involve machinery, and which lighten the burden on women without handing the task over to men. Such innovations are particularly welcome, because although women are often over-burdened, Judy Bryson argues, they 'take considerable pride in their agricultural work', and lose independence if all jobs outside the home disappear through mechanization.[10]

The reason men are attracted to mechanized jobs may be to do with the higher productivity and earnings associated with them, but seems also to be partly due to the way machines convey prestige. The modern male takes pride in being mechanically minded. Furthermore, as with the ancient Greeks whose men worked forges or steered ships while the women worked at the loom or hand-mill, there seem to be very particular values involved in controlling inanimate force – even in the undramatic setting of a ploughed field – and these recall the virtuosity values discussed in the previous chapter. So in modern Africa, as in ancient Greece, there appear to be three kinds of values involved in the practice of technology – firstly, those stressing virtuosity; secondly, economic values; and thirdly, values reflecting the work traditionally done by women.

Users and the management of process

In previous chapters, values and world views have been discussed mainly in terms of the attitudes of economists and more technically-minded commentators. Now we find that women's work roles imply a third point of view, and one which may come closer to the basic needs approach encountered in the Kerala case-study of chapter 4 than to anything else so far discussed.

In order to see how these three viewpoints are inter-related, it is useful to turn back to the 'map' of technology-practice presented as figure 6. This illustrates how we commonly think about technology in terms of an 'expert sphere' of interest, forgetting the activities and experience of 'users'. The expert sphere includes research, design, all industrial activity, and also the professional interests of engineers and medical men. Even though many user-related factors are neglected, this is still too large a field for a comprehensive view to be easy. Most discussion therefore tends to be biased either toward economics and the management of technology (i.e. topics on the right-hand side of figure 6), or alternatively toward scientific knowledge, innovation and technique (i.e. the left-hand part of the figure).

Thus views on the practice of technology tend to be polarized between economic and technical views, and in chapter 4 we see how this leads to very different beliefs about resources; similarly, in chapter 5, we noted two viewpoints about imperatives and goals in technological development. Table 5 summarizes and amplifies these contrasts, and introduces a third viewpoint representing the attitudes and needs of users. This is associated with what I shall refer to as 'user' or 'need values', and 'need-oriented' goals.

It is notable that the traditional division of labour between men and

TABLE 5   Three sets of values involved in the practice of technology

	Virtuosity values	Economic values	User or need values
Exemplars	adventurers (Odysseus), smiths, warriors	merchants, working men	women (Athene, Penelope)
Applications	tractor driving	cash cropping	gardening
	high technology (aerospace, weapons)	production engineering	craft work, appropriate technology
	——————	food technology	cooking, handmilling
	heart transplant surgery	drug manufacture	childcare, primary health work, nursing
Priorities	pursuit of the technically sweet	pursuit of profit	maintenance, subsistence
	mastering natural forces	managing a workforce	care for people, care for nature
	extending frontiers	economic growth	stability
View of technology	construction for prestige value	construction, production for exchange value	management of process: use value
Typical evidence of 'progress'	improving performance (figures 3 and 5)	increasing GNP	falling infant mortality (figure 7)
Attitude to risk	risk as challenge; offset by fixes	risk balanced by potential gain	risk avoidance and prevention
Views of creativity	innovative, adventuring, unrestrained	equated with enterprise	tempered by responsibility
Cross references: Table 3 Figure 6	technical fix expert sphere	economics expert sphere	bio-economics user sphere

women casts men as the makers of tools and equipment, thereby giving them a great interest in the 'expert sphere' of technology, while women are often most directly concerned with the end-use of equipment or energy, and with meeting basic needs. Thus women tend to experience technology less as making things and more in terms of the 'management of process' (p. 68), leading to a very distinctive outlook. The importance of this is not usually recognized because of the habit of regarding women's traditional roles as service activity, subsidiary to the more serious business of wealth creation.

One way of illustrating a distinctive basic-needs approach could be to refer back to the Kerala case-study, because in Kerala (and Sri Lanka), the openness of politics, and the tendency of governments to be prisoners of the popular will, means that economists and economic values are less dominant in policy-making than elsewhere, and user values and basic needs carry more weight. For example, it has been said of both states that agricultural development during the last two decades has been nutrition-oriented rather than economically-oriented; output of basic food crops has certainly increased impressively. However, we may also understand user values better by giving further attention to the traditional roles of women, and that is what will be done here.

Although economic values are dominant in many aspects of modern life, they are not as comprehensive as some people think. Economists can certainly tell us what it cost to land a man on the moon in 1969, but they cannot say whether this was worth it. That is judged on a different basis, and according to virtuosity values. Economists are equally at a loss when assessing the unpaid work of women in the home, again because different values are involved – need or user values. Thus a proportion of women's work is usually entirely omitted from overall estimates of output, such as gross national product (GNP); 'domestic work is not considered as "real" work because it has only private use value but no exchange value'.[11]

The contradictions implied by this become particularly striking when an activity like grain milling is considered. If wages are paid to operatives in a mechanized mill, the economist counts milling as productive work. When it is done by hand in the home to meet family needs, however, he relegates it to subordinate status as housework. Similarly, the washing and cooking done by housewives may not count as production, though the comparable output of commercial laundries and bakeries does.

Even more remarkable anomalies arise when economists study the division of labour between men and women, and conclude that 'labour is rationally allocated in terms of maximizing income'. As Ingrid Palmer points out,[12] such judgements invariably ignore a wide range of factors, including the special problems of pregnancy and breast-feeding. They ignore the fact that 'bodily resources and . . . bodily needs' of individuals may be sacrificed to maintain production. Thus farm work in some African countries may be kept going at the expense of babies, who are born under-weight and are inadequately breast-fed because of the time and energy mothers are devoting to work in the fields. In such instances, if economic values suggest that the division of labour between the sexes is appropriate and right, then they are in clear conflict with all values that have to do with human need.

'Technology' like 'economics' is a term conventionally defined by men to indicate a range of activities in which they happen to be interested. One academic course on technology that I teach but do not control has a section entitled 'food' which deals with mechanized agriculture and fertilizer manufacture in great detail, but says nothing about cooking. Yet any rational definition of technology such as the one given in chapter 1 would encompass cooking just as readily as it encompasses engineering. After all, cooking involves the application of 'organized knowledge to practical tasks', and it involves 'people and machines', even if the latter amount only to stoves, mixers and kitchen scales.

Nearly all women's work, indeed, falls within the usual definition of technology. What excludes it from recognition is not only the simplicity of the equipment used, but the fact that it implies a different concept of what technology is about. Construction and the conquest of nature are not glorified, and there is little to notice in the way of technological virtuosity. Instead, technique is applied to the management of natural processes of both growth and decay. Child-care, vegetable-growing, bread-making and dairy work all depend on the fostering of growth; other work done by women, ranging from cleaning, hygiene and home maintenance to nursing and the care of the elderly, concerns the management of inevitable processes of decay, and relates to the broader concepts of conservation and prevention discussed in chapter 3.

Appreciation of process in this sense partly depends on accepting and working with nature rather than trying to conquer it, and is a neglected concept in conventional technology. Thus Joan Rothschild has every reason to claim that a 'feminist perspective' can help create a 'soft technology future' where such values as 'harmony with nature . . .

and non-exploitation become integral to technological development'. She is right to stress a feminist view as countering the male interest in dominating nature, and in 'pursuit of . . . "rational efficiency" to the point of irrationality'.[13] What is still lacking, though, is the solid framework for analysis that could come from the concept of technology as the management of process if this were developed using appropriate scientific ideas (e.g. from thermodynamics – chapter 4).

If a modified concept of technology were developed in this way, we could perhaps put the case for recruiting more women into, say, engineering in a very different light. It is a strong case, not only for reasons of equality and (as is also argued) to make up the shortfall in male recruitment, but also because it could force us to recognize that engineering may itself be in need of reform: its practice may incorporate values that alienate men as well as women. In the United States in 1978, only one engineer in 100 was a woman, though there were many more women coming through the engineering schools.[14] In Britain, the figure is only one engineer in 300. It is possible to point to all sorts of barriers in social attitudes, girls' education and the employment policies of firms to account for this imbalance, but it is rare for anybody to identify the problems that also exist in the way engineering is conceived and taught. However, one group of engineers at Warwick University has suggested that part of the problem may be that 'engineering is taught by men and for men'; there is a neglect of 'working and living experience'.[15] Technology-practice not only includes innovation, design and construction, but operation, maintenance and use. When, as so often, engineers under-emphasize the latter, one of the positive contributions of a feminist viewpoint might be in stimulating interest in this aspect.

It may be, however, that if we want to understand women's achievements in technology, we ought not to be content with counting the pitifully small number of women engineers, but we should also recall what was said earlier about declining infant mortality in England and in Kerala. This was partly due to advances in education that gave mothers an increased ability to use simple information about nutrition, hygiene and household medicines. On however basic a level, this is the 'application of scientific and other organized knowledge to practical tasks'. It may therefore be regarded as 'progress' in technology just as much as the improvements in engines represented in figure 5. It is, moreover, very largely the achievement of women: mothers, health educators, nurses and teachers.

Another contribution to technological progress in the area of main-tenance and use is illustrated by the Electrical Association for Women, founded in 1924 by Caroline Haslett, herself an engineer. At a time when her male counterparts in the electricity industry were neglecting sales and service, Haslett encouraged women to take a lead in buying, using and maintaining household electrical appliances. She wrote an *Electrical Handbook for Women,* and her association organized courses for domestic science teachers emphasizing the potential role of elec-tricity in aiding emancipation.[16] The use of electric irons and vacuum cleaners, in particular, expanded rapidly, and domestic use of elec-tricity in Britain during the 1920s and 1930s grew much faster than any other sector of the market. Through this and other consumer organiza-tions, women have contributed more to technical progress than we generally recognize. However, research by Ruth Cowan and others is now showing that 'labour-saving' electrical appliances in the home did not emancipate women in the way that has often been assumed.[17] In America, women still spent 60 or more hours per week on domestic tasks, partly because higher standards of comfort and cleanliness were sought.

What emerges, then, is that women already have important roles in technology – in child-care, in management of process, and in con-sumer work – and there is need to recognize these as well as to encourage more women to move into the expert sphere of practice by taking jobs in engineering. There is danger, though, in stressing what is valuable in women's traditional roles: some employers will discover in this new reasons for confining women to low-paid work in the main-tenance and service departments of industry. Mike Cooley, the trade unionist, has described how his experience of computerization has convinced him that new technology, 'is frequently used to consolidate the unequal pay and opportunities for women . . .'.

Thus women need to fight, 'not only the traditional forms of dis-crimination, but much more sophisticated and scientifically structured ones. There is little indication, even in 1980, that the unions catering for such workers have really understood the nature and scale of this problem.'

Working with union colleagues, Mike Cooley went over some past issues of the principal computer magazines examining advertisements. Of those that illustrated a person with the equipment being publicized, some 82 per cent, showed a woman in some kind of absurd posture which was in no way related to the use of the equipment. There is a continual projection of the view, even in the most serious of journals, that women are to be regarded as ridiculous playthings, just draped around the place for decoration. Not only that, but those who read these journals often do not notice the built-in assumption unless it is pointed out to them. They are conditioned to accept the presence of women in the servicing role . . .

A profound contribution that could be made toward creativity in science and technology would be to encourage the involvement of women in this field at all levels. Not, I must add, as imitation men, copying all the absurdness of men, but to challenge and counteract the male values built into the technology.[18]

Tempered creativity

To talk about the new insights that women might bring to technology could be merely a pious hope unless we appreciate how experience and values interact. Sociologists have sometimes noted that many people divide up experience between separate compartments, and apply different values to each. Thus an individual may operate with one set of values when at work, and quite a different set when at home with the family. There is the danger, then, that a woman engineer would adopt the conventional engineer's mixture of virtuosity and economic values in her professional role, and apply a more need-oriented approach only at home. Stephen Cotgrove suggests that it is as if people leaf through a 'gazetteer' of values when they move into a new area of experience.[19] The challenge to both men and women, therefore, is not just to take on new roles, but to break down some of the compartments that divide up attitudes and life, and add more cross-references to the gazetteer.

What is at issue when Cooley and others speak about 'female values' is not that women and men are inherently tied to opposed ethical perceptions. We are all of us human beings first and foremost, capable of sharing perceptions and experiences to a very high degree. In the last resort, there are only human values, not separate male and female ones. But the traditional division of labour between men and women has restricted many women to a rather narrow range of experience concerned with home and family. It has therefore kept their perceptions focused on only one or two pages in Cotgrove's gazetteer – and these are pages which men have tended to neglect.

We can see what may be involved by referring again to agriculture in Africa. One of the few agricultural experts who has advocated that more technical assistance should be given to women farmers added to his recommendation a comment that: 'The women do seem to have a greater sense of responsibility generally.'[20] In charge of children, the sick, the old and of providing food for all, how can a woman but be responsible in her view of technical innovations relevant to meeting human needs?

At an opposite extreme from rural Africa, we may consider the thirty-one women who were working as research scientists at Cambridge University in 1934, and who included a future winner of the Nobel Prize for Chemistry (Dorothy Crowfoot Hodgkin, in 1964). A brief account of their activities notes that while many male colleagues devoted themselves to 'pure' science, the women were much more willing to concern themselves with social issues related to science, such as malnutrition among the unemployed in Britain, and opposition to the militarization of science. Two also stood as Labour Party candidates in local elections. And one of their husbands, C. H. Waddington,[21] remarked that the women did much to awaken a whole group at Cambridge to the question of social responsibility in science about which he later wrote.

These women may well be untypical – they were certainly uniquely situated. But there are some issues, such as peace and disarmament, where women are fairly frequently seen to be giving a lead. What lies behind this, I suggest, is a particular sense of responsibility arising from the immediacy with which they experience human need. Values are rooted in experience.

Men sometimes do work which gives them similar experience, but many professional jobs, and much wage labour, has the effect of detaching them to some degree from family life without offering any other such close contacts with people. In many of the African countries mentioned, some men have paid jobs in distant towns which take them away from their homes for long periods. In Cambridge, the idea of pure science as an abstract, detached pursuit, distanced some of the men from responsibility.

An opposite situation can be seen in the situation of many craftsmen as they once existed in Britain. Such men lived in the villages they served, and the people for whom they built or repaired farm implements or household goods were often immediate neighbours with whom they had close relationships. Thus they saw the products they made in use, discussed them with the users, and repaired them as necessary. Any impulse to be inventive would always have to take second place to neighbours' needs, and any faults in workmanship would be immediately apparent.

The work of housewives and craftsmen is comparable in this respect, and has other similarities. Cooking and dressmaking are practised mainly as craft technologies. They depend on knowledge gained by experience and on personal judgement more than on theoretical formulations. Accounts of village craftsmen stress that much of their work was routine maintenance and repair, and the same is true of housewives, who spend much time cleaning, washing, altering clothes, and keeping the larder stocked.

The importance of the concept of maintenance has already been amply stressed, along with the fact that it is to engineering construction and the technical-fix approach that prestige accrues. Attention to maintenance work involves values of a different order – care and responsibility, discernment and personal involvement. In maintaining a water supply, we have seen, an engineer may have to forget about status and promotion, for the work often goes unnoticed. In maintaining a motorcycle, you must 'have some feeling for the quality of the work', and a 'sense of what's good. *That* is what carries you forward.'[22]

These caring attitudes will be stronger when affection or personal concern is also involved. One thinks of a mother cooking for her family – a nurse giving an injection to a patient – or, indeed, a craftsman repairing a wagon for a close neighbour. In every instance of this sort, care for the quality of the work is reinforced by care for the person who will benefit. Other values might be involved as well. Child-care involves protectiveness and love as well as the practicalities of hygiene and nutrition. How much more than a craftsman's care may be brought to these activities if they are carried out with love?

Part of the modern problem with technology may simply be that many people have become detached from direct responsibilities of these kinds; but there are also historical changes which may have reinforced this alienating trend. The ideal of pure science and its influence on technology is one factor. Another, going further back, was the displacement of craftsmen by professionals in technology. A parallel development was the emergence of a way of looking at the world in terms of mechanical models. The solar system was regarded as being somewhat like a clock, and living bodies were thought of as assemblies of levers, springs and pumps. Such analogies encouraged a rather insensitive view of the world in which men of science focused on those parts that would fit their models and could be tinkered with.

During the scientific revolution of the seventeenth century, this mechanical philosophy becomes quite explicit. It was promoted by people who took a relatively hard-headed, mathematical approach to engineering, and opposed by those who understood craft methods better, and who were more concerned about the use of knowledge and skill for human welfare.[23] Significantly, one historian has also argued that the scientific revolution was, in some respects, a setback for women, and that the mechanical philosophy was hostile to women's values.[24] Looking back on these and other historical developments, one may conclude with René Dubos that 'the dangers of technology' do not come from complexities that make it incapable of social control, but rather from our values and world view, and 'from man's acceptance that he must conform to technological imperatives' instead of striving for 'true human values'.[25]

Conforming to technological imperatives, we have seen, means pursuing virtuosity and innovative creativity. In some respects, these seem admirable aims, but speaking of what they actually mean for product design in technology, Victor Papanek describes a 'cancerous growth' which has spread from the arts and can be seen most clearly as 'the creative individual expressing himself egocentrically at the expense of . . . the consumer'.[26] In other respects, conforming to technological imperatives means following innovative paths wherever they lead, almost without restraint.

Women in their traditional roles and craftsmen with their social obligations always had to show their creativity in less egotistical ways; and their achievements are given rather limited recognition because, in technical and artistic terms, they were restrained in their originality by responsibility. Rachel Maines says that 'female culture is documented almost exclusively in creative forms', but these are 'rationalized' as the production of rather mundane, useful goods.[27] Such things as the crochet work on a child's bonnet, the cable stitch in a knitted pullover, or the icing on a son's birthday cake are, in conventional terms, of little artistic or technical interest, and are regarded merely as decoration. Yet when men paint pictures to hang in art galleries or the houses of the rich, these are claimed to have great significance as art.

Just as I have found it necessary to point out that cooking falls within the definition of technology, and that declining infant mortality can be seen as 'technological progress', so others have found it necessary to assert the validity of women's art forms. Miriam Schapiro has done this by taking elements from work done by women in dressmaking, embroidery, and table decorations, and building them into pictures of a sort that art galleries will hang. [28]

Lack of recognition seems to be a general problem wherever creativity is so tempered by responsibility that innovation is unobtrusive or design does not conform to accepted 'professional' standards. Not only has craft technology often been under-valued, but reform movements in technology, which assert user values rather than innovative creativity or economic growth have faced the same problems. One thinks particularly of the nineteenth century movements for sanitary reform or for the recovery of craft skills, and of the modern movement for appropriate technology. In the early years of the latter, professionals regularly commented that its methods were inconsistent with doing a proper engineering job, and Third World governments said that they did not want second best technology.

When the sanitary reforms of the last century made careful tests on the design of sewers and began to install improved drainage in London, the established engineers of the time refused to recognize that improvements had been made. One historian who describes the obstacles they put in the way of the reformers notes that some of the engineers involved became folk heroes in their own lifetimes through their exploits in railroad building, but 'showed a swashbuckling disdain for the social evils around them'. [29] Significantly, where such engineers were in charge of sanitation schemes, these became yet another opportunity for displays of technological virtuosity. London's main sewage pumping station at Crossness had a vast engine house with cathedral-like gothic arcading. The engineer of Glasgow waterworks described its structures as surpassing 'the greatest of the Nine Famous Aqueducts which fed the City of Rome'. It would 'remain perfect for ages . . . indestructable as the hills through which it has been carried'. [30]

Similar attitudes, transferred to developing countries a century later, produced many over-elaborate schemes for water and power supply. As Michael Ionides put it, 'civil engineers regarded a river as a sort of challenge: how many dams could they build on it, how great a command area could be created, and so on'. But where needs were greatest, which was often 'outside big river systems', the engineers 'had virtually nothing to offer'. [31]

Ethical disciplines

Awareness of such attitudes, and their effect in diverting technology away from the service of human need led Michael Ionides to join E. F. Schumacher in the effort to develop intermediate technologies that were more appropriate to meeting needs. One early project dealt with rainwater catchment systems that could be used in places where big dams were inapplicable. But two points arise that go far beyond individual inventions such as that. One is the nature of the organizations which control technology and which may prevent appropriate forms of technology being used. This is discussed in chapters 7 and 8. The other point is a question of ethics and the individual technologist. One suggestion is that there should be a Hippocratic oath in engineering like that in medicine. There is a difficulty about this, however, in that if such an oath were to have meaning, it would need to take an attitude to the armaments and related industries in which perhaps a third of all British and American engineers are employed.

Another problem is that the Hippocratic oath in medicine does little to make doctors prefer the humbler branches of community medicine to prestigious hospital jobs or work of high technical interest in, say, transplant surgery. For the individual, the effort to seek a balance between virtuosity values and user or need values involves more than giving assent to an ethical principle; it requires also a discipline and a process of personal ethical development. Meredith Thring attempts to provoke thought about this by linking his proposal for an engineering Hippocratic oath with a 'moral spectrum of engineering' which accords low value to weapons development and 'cosmetic machines', and high value to technologies that enhance possibilities for the fulfilment of human potential.[32]

Others refer to the discipline that is required in any individual realignment of values by using words such as 'dialectic' or 'reversal', both of which imply an interaction between different sets of values. For example, Joan Rothschild talks about a 'dialectical vision of . . . technology' in which the conventional values of male and female interact in a new creativity.

Robert Chambers uses the idea of reversal in discussing the attitude required of all professionals – not just technologists – whose work is concerned with basic needs in rural developing communities. He points out that if one operates with conventional values, many aspects of life in such communities will be overlooked. One will see irrigation as canal systems watering hundreds of hectares, and forget the majority of farmers who may be irrigating from wells on their own land. One will think of research institutes or government departments in terms of the well-known men administering recognized activities, forgetting the 'few departments where women are numerous and sometimes pre-dominant'. But it will be the latter, despite their low status, which deal with the most vital, need-oriented subjects: 'nutrition, home or domestic science, childcare, handicrafts, and women themselves'.[33]

The kind of reversal required is summarized for Chambers by a Biblical quotation: 'the last shall be first'. Its application here is that need-oriented work must systematically give first place to those things that convention leaves until last: maintenance rather than construction, nutrition and child care rather than engineering, and the question of what people need, not what professionals can supply (compare chapter 3). Above all, putting the last first means a new professionalism, more aware that 'the technical and neutral appearance of many decisions and actions is deceptive', and that some people gain and some lose by almost every project.

A reversal is also entailed in the new ending that Goethe gave to the old story of Faust, the man of learning who bargained with the Devil. No amount of knowledge, worldly experience or magical power could give Faust satisfaction; but in the end, by devoting himself 'to a socially useful purpose' – land drainage and reclamation – Faust did find contentment. Goethe's play is complex, and for some commentators, its ending is unconvincing.[34] What matters here, though, is not the meaning Goethe intended, but new meanings the story has gained for participants in modern high technology. For one of them, the legendary Faust and the great engineer I. K. Brunel stand together as symbols of what we most admire as well as fear about technology. For another, thinking of comradeship experienced during intensive design work for a nuclear reactor, it is the co-operative aspect of Faust's drainage scheme which is significant.[35] For a third writer, an engineer, it is sufficient to point out that Faust only found satisfaction when he turned from the pursuit of knowledge and magical power to more ordinary work of immediate social benefit. Where such human purpose is lacking, these men indicate that technical endeavour can lead to a sense of emptiness; they then speak of work in high technology as 'playing with toys'.[36]

Few professionals would admit to this, but in other circumstances, and for other men, one may see how vital links with social purpose are broken. In one of the African communities mentioned previously where women do the most essential work in agriculture, somebody openly said: 'a man is a worthless thing . . . what work can a man do? A woman bears a child, then takes a hoe, goes to the field, and . . . feeds the child [through her work] . . . Men only build houses.'[37]

Perhaps in western man, though lack of social purpose and of need-oriented work is not recognized as a problem, there may still be a feeling of emptiness. Division of labour may have gone so far that many people – men especially – feel disconnected and alienated from the ultimate goals of their work. And that may partly be what leads individuals to seek satisfaction in exaggerated expressions of virtuosity, like Faust with magical power or the adventuring Odysseus. Much professional culture may be a disguise for this sense of emptiness, devoted as so much of it is to building up a notion of the social importance of the profession. But as in Faust's final fulfilment, there is still an opportunity to find peace of mind in practical tasks, 'as simple as sharpening a kitchen knife or sewing a dress or mending a broken chair'.[38]

In the technologists' version of the Faust story one may pick out three stages of ethical development. Firstly, the reversal: Faust's turning away from playing with magical power to accept the discipline of ordinary work; secondly, his work of direct service to the community, almost a work of charity; and finally, the fulfilment he ultimately reaches – innocent contentment.

Francis Bacon, a central figure in the scientific revolution, referred to an earlier version of the Faust legend in warning that dangers as well as benefits could come from the growing knowledge and skill he saw around him. He proposed ethical ideals concerning the proper use of science, which successive generations have reinterpreted to meet the needs of their own times. For the 1980s, one cannot do better than use an interpretation due to Jerry Ravetz, in which the same three stages of ethical development are distinguished: 'discipline is the preparation, charity the way, innocence the end'.[39]

The emphasis on discipline came for Bacon as a reaction to vanity in intellectual pursuits. The term 'charity' is from the same root as 'caring', but also reflects Bacon's deep reading of the Bible, where it means 'love'. For Bacon, it is a 'positive ethical teaching . . . charity as practical action for relieving the suffering of individuals'. Men should seek knowledge and practical skill, Bacon said, not 'for pleasure of mind, or for contention, or for superiority to others, or for profit, or fame, or power, or any of these inferior things; but for the benefit and use of life'. As for the future of knowledge-based skill, he prayed that men would 'perfect and govern it in charity'.[40]

The third stage of ethical development, corresponding to Faust's final contentment, is seen as innocence by Ravetz because Bacon was thinking about the 'kingdom of heaven, where-into none may enter except as a little child'. As Bacon well knew, there is a consistent biblical view that this kingdom of the spirit may be found through humble, practical tasks, especially acts of mercy and healing, and through the sharing of bread which 'earth has given and human hands have made'. Bacon's contemporary, George Herbert, wrote about making 'drudgery divine' in sweeping a room. But the peace of mind achieved this way is not just passive contentment; it also involves enhanced awareness of the significance of what is done – one 'sweeps a room as for [God's] laws'. Such work leads to 'awakening', and significantly, when men talk about it they seem most often to have women's work in mind.[41]

Similar ideas may be found within other cultural traditions also. In Sri Lanka, a strikingly similar ethical system has been incorporated into technology-practice by a voluntary organization, the Sarvodaya Shramadana Movement. This organization is engaged in education and health work (including some malaria control)[42] in 1,900 villages throughout the country. Much of its work centres on mothers' groups and farmers' groups; it promotes village technology units and has a training workshop in which youths learn welding and the use of basic machine tools.

Though inspired by some of Gandhi's ideas, the Sarvodaya movement is predominantly Buddhist in outlook, and has monks among its activists. The aim of its work in technology is related to the Buddhist concept of 'Right Action', and the idea of *dana* or giving which this involves. But while to most Buddhists, *dana* means gifts of money to religious causes – food for monks or gifts toward building pagodas[43] – in Sri Lanka, the Sarvodaya Movement stresses *dana* as sharing. This means sharing knowledge through teaching, through health education and in medical work. Most important, though, is the sharing of labour (*Shramadana*), stressed in the movement's name. This takes practical form in work camps where students and villagers together undertake pick-and-shovel soil conservation tasks, or road-building or latrine construction, and join in discussion, education and some meditation also.

The philosophy of all this is formulated in complicated Buddhist terminology and by an elaborate table (much simplified in table 6). What is striking, however, is that the broad outline covers the same general points as Ravetz's interpretation of Bacon. The central concept in both is charity or sharing or *dana* as the motivation for technology. In the Baconian scheme, 'charity' is 'practical action to relieve suffering'. In the Sarvodaya Movement, it is 'compassionate action to remove causes that bring about suffering, fear and grief'.[44] It involves tasks related to irrigation and agriculture, nutrition and health care (table 6).

Many objections can be made to ethical schemes such as this. They can degenerate into empty formality or sentimentality, and like much thinking about social responsibility in science, failure to consider

TABLE 6   A scheme of right action based on giving (*dana*) or sharing, and applicable in the practice of technology

---

Compassionate action to remove causes that bring about suffering, fear and grief

---

Shramadana – sharing labour:
   a to repair and build reservoirs, irrigation canals and wells
   b for soil conservation work and clearing land for agriculture
   c for building roads to villages
   d for building schools, houses and latrines

Buddhidana – sharing knowledge:
   a in education and literacy training
   b to provide a library service
   c in development education and technical training

Bhoodana – sharing land: for cooperative cultivation

Gramadana – community ownership of land:
   a to eliminate landlessness
   b to increase productivity

Waidyadana – sharing health care:
   a through medical aid
   b through health education and work to improve nutrition
   c through primary health care
   d by improving pure water supplies

Dharmadana – sharing self-knowledge – spiritual development

Source: Sarvodaya Movement *Ethos and Work Plan*, Moratuwa, Sarvodaya Press, 1976, part of Annexure 18.

political and institutional factors can be fatal. The Sarvodaya Movement attempts to avoid a party-political stance, but at the same time has evidently helped to reinforce the values of a basic needs economy in Sri Lanka. We have noted a western commentator describing Sri Lankan governments as subjected to a 'tyranny of public opinion' when it comes to need-related policies in agriculture and health care.[45] But since 1977, there has been a degree of economic liberalization, with the establishment of a free trade manufacturing zone. Advocates of this approach aspire to rapid growth on the Singapore model, and as one of the Sarvodaya leaders says, they have sometimes 'outsmithed Adam Smith himself', and have attacked the Sarvodaya Movement as an attempt to 'revert back to a primitive subsistence and feudal economic system'. At the opposite extreme, critics 'who outmarxed Marx . . . saw nothing but capitalism and imperialism in Sarvodaya'.[46]

Opposition of this sort was such that in 1974, the Movement was in danger of being proscribed; however, its achievements are now widely recognized. Tributes are paid by prime ministers, and there is even a proposal 'to humanize and dynamize the bureaucracy' by sending senior officials from government ministries to stay in Sarvodaya work camps. The philosophy may certainly seem vague and even reactionary in its stress on 'traditional values', but critics can be shown 'quantifiable physical results . . . Buses and motor cars drive on the roads we have built . . . Farmers irrigate their paddy fields through channels we have cut . . . Thousands of unskilled hands have been trained to earn a decent living . . . Children study in schools which we have built'.[47] Statistics such as those given earlier in table 4, p. 71 confirm that somebody, if not just Sarvodaya workers, has been very active about such things.

But there remains the difficulty for western readers that spiritual disciplines seem out of keeping with hard, practical technology, even where it has a user or need orientation. When writers on technology seek inspiration in Buddhism – as Pirsig, Schumacher and Touraine have done[48] – they are attacked by those who see this as mere trendiness which seeks to replace rational thinking by a woolly and eclectic mysticism. But for anybody who thinks that sensitivity in the use of technology is important, even within a rationalist frame of reference, Buddhism can present some refreshingly challenging views. It does not demand belief in any god, but rather adds to humanism by offering a philosophy of life on earth in which there is a sense of continuity and natural process. There is also a tendency to think about human creativity more modestly, with less stress on virtuosity. 'Buddhists walk lightly on the earth; westerners feel they must make their mark.'

For me, as for most other westerners with materialist values and an atheistic style of thinking, mystical language is extremely discomforting. However, there are two points in all this which might well be demythologized and applied, and one further point we should at least think about.

First, we may need an ethical, if not a spiritual discipline in technology, not just a Hippocratic precept. This discipline needs to be capable of dealing with conflicts between virtuosity values and need, between expert and user values. There may be clues to such a discipline not only in the Sarvodaya talk of mental 'preparation' for Right Action and material development, but also in the ideas about dialectic and reversal mentioned earlier.

Secondly, there is the suggestion that ideas such as charity, sharing and the meeting of basic needs should be taken much more seriously as goals in technology. One industrial designer notes that there is a difference between design for human need and design for the market, reflecting the conflicts we need to face between user values and economic demand. Designers may have to work mostly for the market, but may still occasionally practice a 'reversal'; and the suggestion is that they might commit a proportion of their thinking to need-related work, 'tithing' a fraction of their ideas 'to the 75 per cent of mankind in need'.[49] Not everybody has the opportunity to do this constructively, but one British industrialist whose firm makes expensive medical equipment spends a few months each year in Africa, training hospital technicians to maintain much more basic equipment.

The final point, which needs thinking about, but which is not for me to pursue, concerns the origins of these values that are applied in technology. Are they, as Marxists tend to say, merely derivative from social structures, employing religious language as a rhetorical disguise for attitudes that reinforce the established bases of political power? Are they, as many other people, including some sociologists, would argue, developed in response to experience, especially during adolescence, as young women and men work out what roles they are to play in life? Or are there in addition, deeper, perhaps instinctive drives to explain the constant building of cathedrals and Eiffel Towers that reach out toward space, and rockets that actually travel there? Even Buddhists share this impulse, as their enthusiasm for pagoda building shows. And why, even in the atheist West, do advocates of high technology feel a reaching out towards an undeified transcendance in this, and speak about spacecraft or civil engineering as 'aspiration to higher things' and 'diagrams of prayer'? We would perhaps be more successful in controlling and coming to terms with modern technology if we better understood the power of these irrational urges, and the failure of most teachings that stress humbler works of service – Buddhist, Christian or humanist – to curb them.

7
Value-conflicts and Institutions

Babels of confusion

Modern man, it often seems, is divided man. There are no universally agreed goals, no wholly comprehensive systems of values: 'the modern mind is divided – in tension'. Again and again there are attempts to resolve the tension by suggesting a rejection of high technology and reversion to a simpler, more rural way of living. But many of the finest achievements of western culture are the products either of high technology or of the virtuosity values that have impelled it. One thinks of the idealistic engineering of medieval cathedrals, the work of Renaissance artist-engineers, the constructions of Brunel and Eiffel, and the marvels of microelectronics or of space exploration. To disown all that would be both Luddite and Philistine. But to assert the importance of meeting basic human needs, and using technology to that end, is an inescapable obligation. To recognize the necessity for an environmental consciousness and concern for conservation is almost equally vital. But those who advocate a rural lifestyle and rejection of modern technology do not have the answer. Neither, at the opposite extreme, do those who advocate 'total (and implicitly totalitarian) materialism . . . Each of these simple choices has failed'.[1]

One of the most sensitive of all writers on engineering, L. T. C. Rolt,[2] has described how conflicts of this sort developed for him during an apprenticeship with a firm of locomotive builders and in a diesel-engine factory. His enjoyment of the work and his interest in things mechanical was wholehearted. His account amply illustrates many facets of what I have called virtuosity values – especially the aesthetic appeal of machines, craftsmen's 'feel' for their work, and the enjoyment of an elemental mobility in the still novel automobiles of the

1920s and 1930s. Yet there were often conflicts with other values. When the locomotive works closed down and craftsmen were thrown out of work, he saw the commercial world 'robbing such men of their only real asset and source of true satisfaction – their skill'. This experience, coupled with what he saw of the smoky squalor of the industrial towns, led him to exclaim that accountants, with their narrow economic values, were 'unable to see that their financial logic made brutal nonsense in human and natural terms'.

Working as an agricultural engineer in the Wiltshire countryside, Rolt found these conflicts lessened. Later, he discovered in the canal system of midland England, 'the one work of engineering which so far from conflicting in any way with the beauties of the natural world positively enhanced them'. Yet he felt increasingly with Thomas Traherne that man's world was, 'a Babel of confusions: invented riches, pomps and vanities'.

The same conflict is a recurrent theme in the literature of the industrial world. America's tradition of pastoral writing vividly portrays the intrusion of other values for which 'the railroad was a favourite emblem', and the locomotive, 'associated with fire, smoke, speed, iron and noise, is the leading symbol.' Leo Marx[3] points to 'tension between the two systems of value', pastoral and industrial, persisting through a century and more. He argues that Thomas Jefferson's ideal for America was a pastoral one; the continent would be a non-industrial land of farmers and husbandmen with the goal of 'sufficiency, not economic growth'. Yet at the same time, Jefferson was 'devoted to the advance of science, technology and the arts'; he enthused over steam engines, and in government did much to create conditions that would favour the development of industry. This 'doubleness of . . . outlook', far from being a handicap, gave him political strength, and is part of Jefferson's continuing significance: he expresses 'decisive contradictions' in our culture. The question of how we deal with these contradictions is central to the ethical discipline we need in technology, as the last chapter suggested. Its bearing on individual behaviour and on the institutions which manage technology is the theme of this chapter.

How do the decisive contradictions arise? Stephen Cotgrove[4] argues that values tend to 'cluster' around different aspects of experience; thus while an industrialist may operate mainly with 'material values', especially in his day-to-day life in the business world, in other circumstances – at home especially – he may turn to quite different pages in his mental 'gazetteer' of values. Nearly all of us do this, and nearly all of us feel acute discomfort and conflict, Cotgrove suggests, when we find values from one page of the gazetteer invading another. He cites situations where art or sex is commercialized as showing how parts of experience normally governed by non-material values may be encroached on by material or economic considerations. In a similar way, when L. T. C. Rolt saw craftsmen unemployed and human skills casually discarded, this for him was an instance where a human attribute was being judged inappropriately on the basis of economic values.

Technology-practice, I have suggested, encompasses a great variety of human experience, technical, organizational and cultural; it encompasses also the contrasts we have noticed between women's and men's experience. Many different clusters of values are associated with this range of experience, not all of them compatible. Thus individuals feel conflict, and society as a whole is periodically divided by controversy about issues relating to technology.

What matters most both for society and for the individual is not necessarily which values come out on top, but how conflicts are handled. Here two strategies make a particularly telling contrast. The first is to make one set of values dominant. Competing claims made by other values may then be subordinated to this master value or set of values. If the conflict cannot be fully resolved in this way, it can usually be kept under control by adopting a compartmentalized style of thinking in which rebel values are kept to a narrowly defined part of life. This leads to a tough-minded, fundamentalist attitude in which few compromises are made.

J. K. Galbraith points out that technological virtuosity is one of the master values of western society and that other goals are separated from it. Thus there is no acceptable way 'to measure the advantages of space achievements against help to the poor . . . the absolute virtue of technological advance is again assumed'.[5]

An opposite strategy, representing a different kind of ethical discipline, is characteristic of people who are ready to live with a situation in which different values pull different ways. Such people are prepared to tolerate ambiguity and look for compromise. The individual with this style of thinking lets a range of values coexist in his mind, and constantly makes cross references to check one against the other. Somebody who is tolerant of ambiguity in this way will not see issues as stark choices between black and white, but between different shades of grey; in politics, he or she will not be attracted by the extremes either of right or left, but will be somewhere near the centre. Critics will say of such people that they are all things to all men, wanting to have everything both ways.

It was such tolerance of ambiguity that was so characteristic of Jefferson's attitude to technology; one cannot cancel out either his ardent devotion to the rural ideal, nor his deep interest in science and technical progress. Deep 'ambiguities lie at the centre of his temperament' and all his commitments involve striking polarities. He admired simple, unworldly, rural lifestyles, but sought high office and cultivated the high arts. The way to understand this, Leo Marx argues, is to see the controlling principle of Jefferson's thought not as 'any fixed image of society. Rather it is dialectical'. It lies in a constant redefinition of his ideal, 'pushing it ahead, so to speak, into an unknown future to adjust to ever-changing circumstances'.

It is precisely this kind of dialectic, I suggest, that we need in thinking about modern technology. We have already encountered it in the discipline of reversal – the practice of periodically turning round conventional attitudes and looking at the world in terms of basic needs and low-status occupations. We have encountered it also in the argument that views of progress in technology should not be fixed in linear images, but subject, like Jefferson's ideal, to constant redefinition and redirection.

However, there is a good deal of evidence that many scientists and engineers have an opposite cast of mind – that they tend to be intolerant of ambiguity. They like to tackle problems which have definite solutions, and feel ill at ease with open-ended questions. 'I'm not fond of debate . . . I prefer analysis', said one leading technologist welcoming a report on a nuclear energy project 'for its lack of ambivalence'.[6]

Similar attitudes have often been noted by educationists[7] among students who show an aptitude for science and technology. These individuals enjoy mathematical problems but very often dislike writing essays, not necessarily because of any lack of literary ability, but primarily because an essay is open-ended. There are no precise rules, and no unambiguous right or wrong answers. This makes sense, because engineers need to avoid open-ended situations in their work. Commitment to being practical and making things work means identifying a viable solution to the problem in hand and concentrating on that. To explore too many alternatives will often mean dissipating effort without getting results. Thus while a good scientist must be able to produce original ideas, a 'good engineer is a person who makes a design that works with as few original ideas as possible'.[8]

What this boils down to is that the most effective engineers are often unusually singleminded, capable of being strongly committed to the task in hand, and able to keep emotional problems from affecting his work by compartmentalizing one broad area of life from another. A remarkable instance of this singleminded approach is seen in the career of Wernher von Braun, who in adolescence dreamed about space travel and rocket engineering, and forty years later was a leading participant in the American programme which placed men on the moon. At every decisive point in his career, from his student days in Berlin, he took the path that would best allow him to pursue his dream. One can admire this commitment, yet feel it showed an almost inhuman disregard for more ordinary concerns. In the 1930s, he worked on rockets for the German army, seeing this as 'a stepping stone' into space. In 1942, after the initial flight tests of V-2 rockets, he almost forgot the war in his enthusiasm for the first excursion into space by a man-made object. He was reported as saying that the V-2 was 'not intended as a weapon of war',[9] and as a consequence, was briefly arrested by the Gestapo. Even as Germany faced final defeat, he was calculating that his chances of continuing with rocket research would be best if his team were captured by American forces rather than by the British or Russians, and managed to arrange this outcome. But when he got to the United States, he was disappointed that the Americans were interested only in rockets as weapons and found that he had to 'evangelize' for the idea of space exploration as a real possibility.

The technocratic master value

Wernher von Braun's vision of how man could venture into space was one about which many people could feel some excitement. The vision of the nuclear scientists was more esoteric but no less intense, and many of them too have participated in research on weapons as a 'stepping stone' towards the realization of technical ideals. Herbert York has described how he was 'strongly motivated and inspired . . . to participate in the hydrogen bomb programme' of the early 1950s, not least because of its 'scientific and technological challenge'. But in York's later, regretful opinion, the effect of the programme was that 'the last good opportunity to base American foreign policy on something better than weapons of mass destruction' was missed.[10]

Technical idealism also had a role in Edward Teller's opposition to the nuclear Test Ban Treaty of 1963. He was partly motivated by 'a passionate desire to explore to the end the thermonuclear technology that he had pioneered'.[11] To do this, experimental explosions would continue to be necessary. Others opposed the treaty because it would stop the development of a nuclear-powered space vehicle on which they were working. These were motives mainly of scientific curiosity and enthusiasm for technological virtuosity, and not especially warlike. But the singleness of mind that gives such goals higher priority than a nation's interests or the welfare of mankind can also seem alarming. One has the same feeling about Von Braun: what is one to make of a man who, while his country is still at war, is preparing to offer his services to one of its enemies?

Among those involved in the early phases of the nuclear arms race, the most outstanding exception to all this was Robert Oppenheimer. As enthralled by nuclear science as anybody, he was also deeply troubled about atomic weapons and sought to delay the hydrogen bomb programme while avenues for arms control were explored. Unlike many of the other technologists, his mind was not compartmentalized so that social conscience and technical creativity were kept apart. For the singleminded Teller, this ambiguity in Oppenheimer's attitude and the conflicting values that informed his actions 'appeared . . . confused and complicated. To this extent . . . I would like to see the vital interests of the country in hands which I understand better.'

With Teller and Von Braun we see another explanation of the technological imperative complementing those suggested in chapter 5: it is not just that they were men with enthusiasm for technological virtuosity. It is also that this became a master value to the point of obsession. We are all impressed by technological achievement at times and appreciate the aesthetic qualities and sense of mobility associated with some machines. Virtuosity values may thus be values we all share. What sets these men apart is rather the way they built virtuosity into an overall value system, using an ethical discipline whih preserved their central aims from any sort of compromise.

In previous chapters, I have mentioned fictional characters who illustrate particular values – Odysseus, Faust, Captain Ahab. What they had in common was that they were all on a quest or mission which can be seen as a singleminded pursuit of a narrowly defined goal. None of these can be represented as technologists, but the compulsions and disciplines of much technology-practice seem to reflect the same sense of mission; Von Braun's forty-year pursuit of his space-flight vision was itself outstandingly a quest. Von Braun – or a century earlier, Brunel –

might have made clear his undeviating purpose in Ahab's words: 'The path to my fixed purpose is laid with iron rails, whereon my soul is grooved to run . . . Naught's an obstacle, naught's an angle to the iron way.'[12]

In moderation, these attitudes may lead to a determination and decisiveness that we reluctantly admire. But they may also lead to obsession and irresponsibility, and in Ahab's case, to 'monomania' and near madness. In western society, economic growth is said to be a master value, and this too is something about which one has very mixed feelings. It has in many ways been a goal whose pursuit has brought benefit. But when it leads to a singleness of mind which is willing to 'cut down the last redwood, pollute the most beautiful beaches, invent machines to injure and destroy plant and human life', then one must agree that to have 'only one value is, in human terms, to be mad'.[13] It is not that any particular master value – growth or virtuosity – is wholly mistaken, but simply that by itself it is inadequate and incomplete. A multiplicity of values are a prerequisite for a balanced life.

Thus my arguments in favour of user/need values are not aimed at turning the service of basic human need into a master value and withdrawing moral sanction from high technology. That would merely replace the tough-minded pursuit of virtuosity with the equally un-balanced drives of do-gooders. What is more essential is tolerance of a wide range of values, and a determination to make creative use of the tensions between need-oriented, nature-conserving and virtuosity-related goals.

Such tolerance comes hard to many working technologists, not only because they do not want divergent sets of values distracting them from the job in hand, but also because they are heirs to a conventional wisdom which is designed to minimize ambiguity and the debate it can lead to. Thus the conventional wisdom implicitly encourages the idea of a master value such as economic growth; it encourages an un-ambiguous approach to problem-solving also, frequently favouring a technical fix approach because this may avoid the messy complications of a more human solution, and is often within the capability of a self-sufficient, specialist profession.

Within this conventional wisdom, beliefs about progress are also very clear-cut; it is regarded as unambiguously logical and linear, occupying a single dimension of forward advance. And it is anticipated that future needs will line up with the direction in which technological imperatives are leading. So when experts present forecasts that appear to be dishonestly biased, there is often no dishonesty at all – such projections are a straightforward interpretation of a particular view of progress and its imperatives. For those who are intolerant of ambiguity, there is little room for debate about the future: there is only one way forward and the expert knows best where to look for it.

Thus we see that all the varying aspects of the conventional wisdom described in previous chapters fit together. They form a complex which we can describe as a technocratic value system; they give rise to what is often called a 'technocratic' outlook that is singlemindedly insistent on an unambiguous view of progress, of problem-solving, and of values. The word 'technocratic' is appropriate because this is a world view which leaves very little room for democracy in decisions affecting technology. Any idea about choice of technique (or altered priorities, or public participation in decision-making) introduces a note of un-certainty which is fundamentally unacceptable to those who take this view. To them, there cannot be any rational alternative technologies because there is only one logical path forward. To them, critics of technology are always opponents, never reformers. Yet engineers and other experts need to be continually challenged by reforming critics as a reminder that the virtuosity values which tend to capture their enthusiasms may be in conflict with those of society.

In medicine, for example, the technical interest of highly specialized treatments or operations diverts doctors from more essential but more basic work. One critic argues that, to keep some sort of balance, we must limit the tendency for medicine to become an 'ever more complex technology . . . We must keep it and its advocates, doctors and com-mercial entrepreneurs, under control.'[14] Similar comments about nuclear arms have already been quoted: Zuckerman has called for 'a control of research and development' of a kind, 'which has not existed hitherto'.[15]

Totalitarian institutions

In some branches of technology, the tendency for experts to pursue goals of their own that diverge from the wider aims of society reflects the incompatibility between virtuosity values and other goals – between technology as an end in itself, justified in 'cultural terms', and tech-nology 'as a means to other ends'.[16] The conventional wisdom is that technology chiefly serves economic purposes. Galbraith stresses that technological virtuosity is only a subsidiary goal of industry. To ensure its survival and expansion, every industrial concern must first achieve economic success. More generally, technologists are portrayed as 'servants of power'.[17]

But very large sectors of high technology in America and Europe have escaped the economic constraints normal in private industry, and are to be found in government-supported or subsidized defence, nuclear energy and aerospace industries. In these circumstances, if technologists are servants of power, the situation is nonetheless quite different from industry. By clothing goals related to technological virtuosity in the language of military necessity or political prestige, and by pointing to spin-offs for the economy, it is easier than in industry to influence decision-making. But the history of the arms race and nuclear energy, and the comments of some politicians on the handling of information by scientists and civil servants,[18] suggests that behind the economic and military arguments that serve to disguise their virtuosity-oriented drives, these people sometimes use their specialized knowledge in ways that make them hijackers of power, not its servants. Even in private industry, where this is less likely to occur, the development of some products owes much to the hunches and singleminded backing of an individual staff member, who works for favourable decisions at every stage in its development. Such individuals, referred to as product champions, were identified in one study for some 40 per cent of all innovations examined.[19]

Von Braun and the early nuclear weapons scientists were product champions on a grand scale. But their significance lies also in the new institutions for the management of technology that grew up around the projects they led – institutions whose mission-oriented structures reflected the personal questing sense of these men. The chief examples were, of course, the German centre for military rocket development, opened at Peenemünde in 1936, and the American atomic weapons programme located at Los Alamos. Their descendents in the modern world include NASA, numerous weapons laboratories, and the Atomic Energy Commissions of nations as diverse as the United States and India, France and Argentina.

We saw in chapter 2 how the first industrial revolution originated with an organizational innovation – the factory as an institution for controlling a workforce. The new wave of industrialization which originated just before and during the Second World War also depended on organizational innovation, affecting particularly the way research and development are done. There was a rapid development of existing industrial laboratories and experimental stations as well as the mission-oriented nuclear and aerospace projects.

In these institutions, scientists and engineers had a new prominence, and the question again arises as to whether they were still servants of power. Several interpretations are possible, but many people have noted that the administration of some of the new institutions has become so intertwined with departments of government and the civil service that technical experts have, at the very least, become identified with decision-making power.

However, to understand the role of the technical experts, a distinction between two sorts of power may be helpful. The politician's business is power in society. In itself, this is probably of little interest to most technologists. What they want is not a generalized power over people, but power over specific projects, and power to exclude people from interfering with them. In one psychological test, when aspiring scientists were asked to sketch a street scene, they tended to omit people, preferring lunar landscapes.[20] Similarly, an ideal factory would be an automated one that employed no workers. In other fields, too, the world of the technical idea is often a self-contained one, involving neither lay participation nor co-operation with other experts. Indeed, departmental, specialized interests sometimes seem so strong that experts are prepared to overlook risks to world peace in this interest – or, like the German rocket experts in 1945, to work for any country which would support their projects. It is when their idealized worlds are threatened that technologists are tempted into the manipulation of knowledge and through that, into hijacking power.

Of course, there are many situations where 'people problems' and the user sphere must inevitably intrude upon idealized technical thinking. Then experts may be tempted to believe that technical rationality can still be achieved through use of a systems approach, or by planning on a sufficiently comprehensive scale. During the 1930s and 1940s, scientific socialists and humanists talked enthusiastically about large-scale planning. In one of the most revealing of their comments, C. H. Waddington cited the German autobahns and the Tennessee Valley Authority (TVA) as good examples of how rational planning could be conducted. These examples, he thought, illustrated a trend towards totalitarian organization, and he argued that this was an inevitable and desirable part of technical development.[21]

Waddington published these remarks in 1941, in the middle of a war against fascism. Thus he could not avoid noting that totalitarian systems were getting a bad name. But the objectionable features he freely admitted in Nazi and even Soviet regimes did not seem inevitable. What we needed to do, he argued, was to work out how 'to combine totalitarianism with freedom of thought'. Several other authors during the 1940s commented on the same trend, J.D. Bernal favourably, and George Orwell with vehement criticism. These writings are of significance not only because of the frankness with which they discuss totalitarianism, but also because this was a formative period for the new technological institutions, and pressures of war exacerbated the totalitarian tendency.

Although the origins of these organizations are linked with scientific discovery, they also represent a merging of science and technology and changes in the accepted standards of professional behaviour among scientists and engineers. Teamwork was essential, and so was secrecy. Significantly, however, the habit of secrecy persisted in peacetime, and in non-military programmes. In no country which now has a nuclear programme was the decision to embark on it taken openly and with democratic, parliamentary approval. That was understandable in war. But in the United States, the Atomic Energy Commission (AEC), proceeded throughout its early years with a cloak of secrecy surrounding even civil energy projects. One of its officers, Herbert Marks, warned that if this continued, the atomic energy programme would lose touch with the American social ethos, so that 'when the forces of criticism finally begin to operate with their customary vigour, they will produce drastic upheavals'.[22]

One function of secrecy has been to reinforce linear, mission-oriented thinking by ensuring that ideas, innovations and doubts can only be expressed through the institution's own bureaucratic channels, and not in the press, Congress or Parliament. This protects the central goals of the institution from abiguity or uncertainty by making sure that criticisms or divergent, perhaps irrelevant, inventions are compartmentalized by bureaucratic procedure. The same procedures also ensure that no individual carries unique responsibility, thus encouraging people to feel that it is not incumbent on them to raise questions. All this, of course, goes wrong when doubts or counterproposals are raised at the very top of these organizations – when an Oppenheimer begins to entertain ambiguity about weapons which he simultaneously finds technically sweet and morally repugnant.

One result of this tendency is that errors, once made, are reinforced as often as they are questioned; conversely, technical innovation may be suppressed. The totalitarian bureaucracies of British technology have functioned particularly badly in these respects. In the nuclear power industry, and in decisions about the Concorde airliner, procedures were apparently designed to avoid the consideration of too many points of view; secrecy was enforced to an extent which made it difficult to learn from past mistakes. In addition, David Henderson has noted a peculiar British taste for decorum and administrative tidiness which seeks to avoid duplication – often, just at the point where a duplicated assessment or evaluation could provide an essential cross-check. In British aviation and nuclear programmes, the remarkable result has been that expert advice is obtained only from 'interested parties to a decision . . . decorum precludes any serious attempt to make use of alternative sources of advice'. To counter this trend, Henderson argues for new and independent institutions that would evaluate public projects from non-government points of view.[23]

Institutionalized ambiguity

All the foregoing applies only to one sector of technology-practice. There are many small firms and even academic laboratories which are in no way bureaucratic, and where an open-ended approach to innovation may be found. There are many individual engineers whose outlook is more flexible than those I have described, and there are concerned scientists swimming against the tide. But the compatibility between the singleminded individual with his master value and the bureaucracy with its focal task seems both striking and significant. It poses a question about whether these similarities arise because individuals internalize the values of the institutions within which they live and work, or whether, perhaps, they help to shape institutions in accordance with their own goals.

Sociologists tend to depict the values of the individual as developing to fit him (or her) to the community in which he has to live, and they talk about the way in which values are then used to legitimize institutions. Marxists tend to say that values stem from social and economic conditions, and that the key to change is thus reform in socioeconomic structures.[24]

By contrast, it is possible to see values as fundamental to the individual – as the personal criteria we apply to the world we look out on. Some psychologists and others quote evidence to show that the individual's value system is partly the outcome of his temperament. The kinds of institution he seeks to work for are then those into which his personality traits fit in a congenial way.[25] This gives scope to the view that institutions develop at least partly under the influence of individuals' values and actions.

The emphasis of much that is written about technology is on institutions rather than individuals. That is often valuable, but it is only half the picture. My concern is with the other half, and my approach here is deliberately non-sociological and non-Marxist, not for the sake of disagreeing with these types of analysis, but in order to take some account of very basic personal values that are rarely considered, and to think, very simply, about the ethical disciplines that are entailed in resolving value-conflicts.

So far we have examined mainly the tough, singleminded approach which subordinates everything to a master value. However, there is also the quite different habit of thought described by words such as dialectic or reversal. Thus Thomas Jefferson is said to have derived his vision from conflicting values – pastoral on the one hand, and intellectual and technical on the other – and to have let both inform his actions. Significantly, then, this shaper of democratic institutions, unlike the shapers of totalitarian programmes mentioned earlier, did not have a master value or quest with a fixed goal. Instead, his goals were progressively redefined in the light of events, and as values interacted.[26] This dialectical style of thinking, with its reversals of viewpoint and redefinitions, is precisely the opposite of the singleminded approach, allowing options to open and directions to change instead of seeing progress only in linear terms.

It is apparent then, that both in the thinking of individuals and on broader policy levels, the tough, singleminded approach is basically inflexible. The institutions of free speech, congresses and councils, by contrast, are designed for the very purpose of allowing a dialectical process to take place within society, not only so that people's rights can be safeguarded, but so that continued adjustment of goals is possible.

Although the institutions of free speech encourage a great variety of values to coexist, they depend very much on a common view about how value-conflicts should be dealt with, and in this respect, they depend on a very definite value system. We may describe it, indeed, as a 'democratic' value system, in contrast to the technocratic one mentioned earlier (page 127). The implications of a democratic approach in this context are a stress on diversity, flexibility and participation. The latter does not just mean formal public participation in decision-making (which will be discussed in chapter 9); it also refers to a style of innovative activity in technology whereby new insights arise from the interaction of different interests and ideas (discussed in chapter 8).

Diversity and flexibility may be interpreted in terms of encouraging small firms rather than very large ones. Diversity may also be thought of as favoured by community enterprises, and by regional or municipal forms of public ownership rather than by a centralizing form of nationalization. But equally, flexibility may mean that a nation walks on two legs, with a few large-scale enterprises operating alongside many smaller ones. There should be a corresponding diversity in manufacturing techniques, energy supplies and agriculture. With regard to all these things, the approach to avoid is the one that looks for a single, standardized right answer. Whether standard solutions of this sort are conceived in terms of an all-electric economy, or a narrowly technical green revolution throughout the Third World, they invariably lead to gross distortions.

With a more diverse and flexible approach in technology, a more responsive attitude to public participation and democracy in decision-making ought to be possible. But for decisions to be intelligent as well as democratic, the idea of diversity has to be applied to research, and also to sources of information and advice. The tendency to think that there is always one best answer to any technical problem has led to a strongly entrenched assumption that only one kind of research is possible. Yet it will be argued in chapter 9 that there is a need for much more in the way of public-interest research. This should question – but not displace – the orthodox research carried out in official research laboratories and by industry.

In thinking about these issues, we may well conclude that the problem with control over technology-practice in western society is that no nation is wholly democratic and free; everywhere there are totalitarian institutions which have gained control of large sectors of technological endeavour, and which limit diversity and participation unnecessarily. If the 1940s and the extremities of war provided the circumstances to allow many of these totalitarian technological institutions to take off, the 1950s, and particularly the later years of the Eisenhower administration (1953–61) saw the first recognition at a high political level of the totalitarian threat of the political process.

Eisenhower's final address to the nation as president, given on television in January 1961, is famous for its warning against the military-industrial complex. More specifically, he also warned of the danger 'that public policy could itself become the captive of a scientific-technological elite'. It was with regard to arms control that Eisenhower had experienced the problem most keenly: 'I lay down my official responsibilities in this field with a definite sense of disappointment,' he said.

The day after his broadcast, people were asking whether Eisenhower had turned against science. He emphasized – as he had already done on television – that he was in favour of scientific research, 'and feared only the rising power of military science'.[27] In 1958, after America's first satellite had been put into orbit by a military rocket development under Von Braun, Eisenhower had insisted that NASA be set up as a civilian space agency. It should be entirely open in its work, so as 'to have the fullest cooperation of the scientific community at home and abroad', and to ensure 'that outer space be devoted to peaceful and scientific purposes'. But that is another ideal frustrated: of twelve satellites placed in orbit by NASA in 1980, ten were for the Defense Department.

One way of expressing the issues raised by the technology-based bureaucracies, and by the big multinational companies, is to say that nations which are nominally democratic are finding that large sectors of decision-making have been taken over by totalitarian institutions. Looking at the economic role of these institutions, Ralf Dahrendorf has argued that as economic growth no longer makes sense as a master value, we need to think of an 'improving society' rather than an expanding one, and an 'economy of good husbandry'. As he sees it, the institutions of an improving society have to be 'public, general, and open'. But he is not just arguing 'for the simple reconstruction of representative government as we have known it in the past'. In confrontations with large organizations, we are always likely 'to find elected assemblies on the shorter arm of the lever'. Thus, to balance things up, we need informed and organized publics.[28]

It is striking that such a commentator at least recognizes the problem, but something still needs to be done about it. In Europe, an important lead is being given by the political ecology movement of France and Germany. There is a tendency among its supporters to identify the growing totalitarian sector in western society not just with large corporations or with the military-industrial complex, but more generally with the knowledge-based power of the technocracy.

The institutions responsible for energy supply – especially nuclear energy – are regarded as being some of the strongest technocratic organizations. So within a comprehensive view of knowledge-based power in technology, the political ecologists have found the chief focus for their action in campaigns against nuclear energy. They 'reject the model of society which is implicit in the way nuclear power stations are run and the way they affect the wider society'. By this they mean the totalitarian structure of the nuclear industry, its secrecy, the extreme security arrangements that necessarily surround it, but above all, 'the technocratic power which is imposing an all-nuclear policy [in France] with a minimum of public debate'.[29]

Many people who have opposed nuclear energy are 'defensive protestors', who react to nuclear installations as objects which pose particular threats to health, and specific problems about waste. These are real problems, but are not necessarily worse than health and waste hazards in chemical industries. Thus after the power plant accident at Harrisburg (Pennsylvania), which galvanized defensive protestors everywhere into new emphasis on the dangers of nuclear energy, little was done by the political ecologists to capitalize on the event. Fear alone, they felt, could not usefully add to understanding of the central issue, technocratic power, and the use of risk as a proxy issue might divert attention from it.

In tackling the underlying institutional problems of technology, political ecology is clearly a radical movement, but it rejects the traditional radicalism of the Left, and is out of sympathy with the tough-minded forms of Marxism. On a practical level, there has also been difficulty over tactics: the violence used by some Left groups at anti-nuclear demonstrations in France and Germany has appalled the nonviolent ecologists, who have found it easier to work with feminist groups and one of the French trade unions. They have also worked with groups interested in self-management in industry, and in Brittany they helped with a local plan for energy and technology – *Projet Alter-Breton*.

On a theoretical level, too, there is difficulty in co-operating with the traditional Left so long as this maintains its 'devout faith' in central planning, large-scale organization and 'the scientific and technological revolution' which was supposed 'to bring social progress'.[30] This latter kind of socialism sees its problems in terms of economic institutions and the need to change relations of production, but in carrying out its programmes, it tends actually to reinforce the totalitarian use of knowledge. Some Marxists are beginning to recognize this and argue that more emphasis must now be given to the role of knowledge, and hence to 'cultural revolution . . . directed to the appropriation of . . . the intellectual forces of knowledge and conscious decision'.[31]

Penetrating Marxist jargon is always difficult, but if 'cultural revolution' means challenging the conventional wisdom of the experts, then there may be common ground here not only with political ecology, but with the arguments of this book. If cultural revolution involves fostering dialectic and reversal to bring need or user values into balance with the values of high technology, that again will give us common ground. If it means challenging the organizations that depend on an ideology of linear progress and the technical fix, then that common ground may be extended. I leave these questions open, and go on in the next chapter to explore dialectic and dialogue in the practice of technology.

8
Innovative Dialogue

Two kinds of innovation

Sometimes one hears over-simple distinctions made between high technology and supposedly more appropriate forms. The point usually is that appropriate technology is employed to serve human needs directly, whilst high technology is concerned with high performance and complexity for its own sake; it is motivated by prestige and virtuosity, and seems sometimes only to produce 'toys' for scientists or politicians (p. 113).

Something of this view has been implied by previous pages, but as a warning against taking such distinctions too far, this chapter classifies technology in a different way. It links together real toys produced at the level of appropriate technology (figure 9) and techniques which are certainly concerned with basic needs, but which have all the complexity and sophistication of high technology (figure 8). The diagrams themselves are symbolic.

My concern in coupling together these sharply different examples of innovation – toy automobiles and a futuristic power plant – is to pursue a paradox left unresolved by the previous chapter. Modern technology is nothing if not innovative, yet bureaucracy – especially if it is any sense 'totalitarian' – would hardly seem to provide the right atmosphere for original, inventive thinking. Given the growth of bureaucracy in the modern world, how is it that innovation continues to flourish?

One answer is that a wide range of innovations arise outside bureaucratic institutions – which is partly where figures 8 and 9 come in. Another answer, though, is that some forms of innovative development do prosper within a bureaucratic context. Where institutions seem to have a restrictive effect, or individuals are narrowly single-minded, few radically new ideas may arise, but there can still be impressive and sustained improvements in established techniques. Modifications to equipment can be built up on one another in a systematic, logical way, conforming with the linear view of technological development. And with large resources devoted to research and development, this approach will undoubtedly get results. Wernher von Braun's work on rockets is a classic example.

By contrast, though, it is often pointed out that many of the most significant new ideas in technology have come from small firms and even from individuals working on their own, such as Chester Carlson (inventor of xerography) and Christopher Cockerell (the hovercraft). Certainly, the lone inventor will often need the resources of a large firm to turn invention into marketable innovation, but the key point is that his initial creativity worked best outside bureaucratic limits. It has similarly been argued that the most creative phase in space technology and nuclear energy was characterized by a stimulating interaction between enthusiastic individuals. As the large institutions which manage these technologies grew more bureaucratic, the most enterprising inventors 'were driven out', according to Freeman Dyson. In rocketry and space research, 'professionals have never been willing to give a fair chance to radically new ideas',[1] and several possibilities for inexpensive space vehicles have been neglected by the big bureaucracies. Similarly, in the nuclear energy industry, Dyson claims that opportunities for safer and cleaner nuclear power plant have been, in effect, suppressed. Several unconventional reactor designs 'disappeared and with them any chance of . . . radical improvement beyond our existing systems'.

The discoveries about the structure of matter which made nuclear reactors possible also led indirectly to the invention of the photovoltaic cell, which is capable of turning solar energy into electricity. But as Anthony Tucker points out, it was inevitable that nuclear power received greater attention during the next few war-stricken years. 'What was not inevitable . . . was the way the imbalance created by war became institutionalized.' That did happen, however, with the consequence that we have seen a linear development of nuclear energy systems, and a relative neglect of photovoltaic technology. Even as a power source for artificial satellites, where solar energy has great advantages, the technique was neglected for a time; during the 1960s, NASA spent twenty times more on developing nuclear devices for this purpose than on solar cells.[2]

In a rather similar way, it has often been argued that the institutional structure of automobile manufacture has led to a linear development based on internal combustion engines, and a neglect of alternatives such as electric vehicles or cars with Stirling engines.

In Britain, the behaviour of some nationalized industries in relation to technology has tended towards the totalitarian model, while others have seemed more open and responsive. Among the latter, the National Coal Board (NCB) has pioneered highly efficient coal-burning boiler plant based on the concept of fluidized bed combustion. In response to trade union pressure for better safety standards, it has developed new mine machinery whose performance has earned some export successes as well as use in British mines. And responding to the needs of a city council, the NCB has helped develop a district heating scheme at Nottingham which actually burns garbage as fuel, with only relatively small quantities of coal. Thus the NCB has been innovative on several fronts, partly as a result of fruitful interaction with other organizations.

By contrast, the Central Electricity Generating Board (CEGB) has been very resistant to innovations marginal to its main interests. It has pursued a linear development of very large power plants linked together by one of the world's largest grid systems. This has allowed electricity production to be optimized using high-performance turbines operated in 'merit order', but at the cost of inefficiency in consumption of primary energy, and arguably a neglect of responsibilities regarding pollution (and especially 'acid rain'[3]).

For a different approach to these issues, one may turn to the Midlands Electricity Board, whose business is chiefly to sell power drawn from the CEGB national grid rather than to generate electricity on its own account. However, it branched out in 1980 by building a small power plant of its own at Hereford. This uses the principle of cogeneration, whereby reject heat from electricity production generates the steam and hot water required to run processes in two nearby factories. This system is common in Europe and the United States, but has been discouraged by the CEGB. In its version of the idea, Midlands Electricity has demonstrated that small, flexible schemes can be more economical than the conventional wisdom assumes, especially if diesel generators are used. Looking to the future, when oil may become scarce or costly, an outline of other options envisages using coal, with a liquefaction process (figure 8). Solid residues from this could then be combined with garbage in fueling a fluidized bed boiler to supplement the heat output. The diagram shows a supply of crushed limestone to the

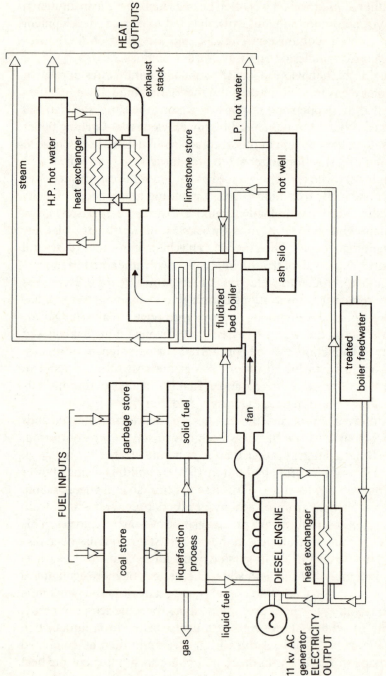

FIGURE 8 *Interactive innovation as it might occur in the future development of cogeneration (or combined heat and power): a diagram indicating long-term options opened up by the Midlands Electricity Board's work*

Source: G. T. Shepherd, 'Combined heat and power', paper given to the Parliamentary Liaison Group for Alternative Energy Strategies, House of Commons, London, 17 March 1980.

fluidized bed to ensure that the sulphur content of the fuel is retained in the ash rather than causing atmospheric pollution.

Figure 8 also shows how a flexible interaction with customers' energy requirements is possible; the plant can be designed to supply heat to factories in the form of steam, pressurized (HP) hot water at a high temperature, and/or low pressure (LP) hot water at temperatures suitable for central heating. Thus the concept developed by Midlands Electricity is capable of flexible development in response to customer demands and environmental constraints. Such responsiveness may, indeed, be said to carry the idea of a dialectical approach into a practical context.

Flexibility of this sort is exactly what the big bureaucracies are bad at, and the example shows what initiatives may be taken by a regional Board which has close contact with its customers. Yet it is still sometimes suggested that the British electricity industry should be further centralized. It is said that government policy would be more readily formulated if the industry spoke with a single voice. But this is the very reverse of what is required. If there is to be a real dialectic on the policy level, we need 'multiple voices', and 'multiple public views'.[4]

When it comes to the development of renewable energy, there has again been little response in Britain to what customers can use, or what British firms might sell. Almost the only official interest has been in prospects for large-scale electricity production by the CEGB. Thus while the Japanese are selling small wave-powered devices for powering lamps in lighthouses and navigation buoys, and while firms such as Lockheed, General Electric, and Saab-Scania are developing commercial wind-energy machines, the British government's chief scientific adviser on energy has reportedly dismissed any similar research with the comment that 'any fool can build a windmill'. What interests him is real engineering, and what the CEGB wants is only large-scale power. This is a virtuosity-oriented attitude if ever there was one, and as a British commentator observed: 'We have a history of vaulting post-war engineering ambition combined with poor manufacturing sense'.[5]

Thus the smaller and more realistic wind and wave-power systems have joined the long list of commercial opportunities that Britain has missed, along with the manufacture of plant for cogeneration (where there are good export markets). In such branches of technology, it is easy to feel that there has been a fairly systematic suppression of innovation in Britain. But it is not only public bodies that are guilty of this. Many instances have been documented from private industry where suggestions made by freelance designers or trade unions have been rejected partly because the ideas do not come through the proper bureaucratic channels. Examples include designs for heat pumps, aircraft passenger seating, washing machines, and telephone equipment, many of which were displayed in London's Design Centre in October 1981. Other instances relate to a new type of replacement valve for use in heart surgery, and a novel petrol-electric road vehicle.[6] All these innovations, rejected by British companies within the last decade, have been taken up by firms in Germany, Italy, Japan or America.

British ineptitude in dealing with novel design or unorthodox invention compares sharply with sustained British progress in a few narrowly defined areas of high technology, such as the design of military aircraft and of very large power stations. The contrast brings out clearly that two kinds of innovation are involved: large bureaucracies, I have already argued, tend to be good at *linear innovation* along established paths. But there is also the quite different type of innovation which bureaucracies tend to suppress and which much of British industry discourages. This depends on the imagination of the creative individual, on interaction among enthusiastic scientists or technicians, and often on interaction between experts and users, designers and potential clients. I shall refer to this as *interactive innovation*.

Cultural exchanges

Sometimes interactive innovation originates in specialist enthusiasm and the exchange of technical information. Sometimes, however, conflicts of values are also involved, and innovation can be seen as the outcome of a dialectic such as the previous chapter described. Transactions at all these levels are most clearly seen in the technological relationships between markedly different cultures: between Japan and Europe, for example, or in western technical assistance to Africa. In some instances, techniques and equipment are simply transferred to a new cultural setting and an attempt is made to impose them where they do not fit. In other instances, however, an innovative process occurs, and techniques or tools are transformed by being incorporated into the recipient culture.

When white men first penetrated into the arctic areas of North America, they were few in number and needed to learn from local people the techniques of shelter, clothing, hunting and travel that were essential for survival in a harsh environment. Thus when white men introduced new equipment – notably firearms – and began to trade industrial products for fox furs and beaver pelts, there was a two-way exchange of technical information with the local Dene (Indian) people, and later with the Eskimos. Local people were not suddenly overwhelmed by the new culture, but were often able to make organizational innovations at their own pace in order to fit the hunting of fur-bearing animals and the associated use of guns and new traps into an evolving lifestyle. The process continued through the last century into this, and as we have seen, it has extended to the point where snowmobiles became part of the local scene. It would be quite wrong to regard this as an idyll, or free of the exploitation that goes with commercial development, but it is also wrong to ignore the positive aspect: many communities were able to incorporate imported technology into their culture by a process of interactive innovation. Thus, in order to use snowmobiles efficiently, Dene and Eskimo people had to invent servicing and maintenance procedures suited to local conditions, where were more rigorous than the designers of the machines had envisaged (chapter 1); they had to adopt a new approach to planning their journeys relative to fuel supplies and the crossing of frozen lakes in seasons when a vehicle heavier than a dog-sledge might not always be secure.

Meanwhile, white men learned in a more technically-oriented way from local experience of the strength of ice alloys, and from Eskimo protective clothing. The latter included waterproof garments for use when fishing from kayaks, and slit goggles or visors to give protection from the glare of sun on water or ice. Their usual winter clothing had better insulating properties than that used by many arctic explorers, even now. Evidence that Europeans have learned from this is provided by the word anorak, which has passed into English from the Greenland Eskimo language.

Those who have studied the specialized environmental knowledge possessed by non-European peoples are sometimes tempted to misrepresent it. One author, referring to arctic clothing, described the Eskimos as 'the great pioneers of micro-climatological bioengineering'. This is inappropriate, because Eskimos clearly do not work with engineering concepts. Theirs is a form of craft knowledge, based on craft technology, and the rather ugly term indigenous technical knowledge (ITK) has served better in recent publications to describe what is involved.[7]

Such compromise as may once have existed in the Arctic between white man, Dene and Eskimo has been damaged in recent decades by a decline in the fur trade, by military radar construction work, and by oil drilling. In Canada's Mackenzie Valley, many of the most critical issues came to a head when oil and gas companies proposed to build a gas pipeline along the valley and southwards into the United States. The Canadian government appointed a judge, Thomas Berger, to conduct an inquiry into the wider implications of the pipeline, and the careful way in which he exposed the conflicts of values involved is a model of its kind. Firstly, he pointed to the (virtuosity-oriented) 'frontier values' of the dominant North American civilization; then there were the values of the existing Dene and Eskimo populations, who saw the Arctic as their homeland, and who feared the social impact and damage to hunting that the pipeline might bring. Need-oriented values were part of this, but more than that, there was a sense of identification with the land as a source of identity as well as of subsistence. Finally, there were the nature-conserving 'values of the wilderness' held by many of the white community who wished to protect the wildlife and the unspoiled beauty of the landscape.[8]

One argument which Judge Berger had to face was that the traditional lifestyle was already dying because of the acquisition by local people of modern rifles and snowmobiles. This had led some observers to argue that local people were willingly committing themselves to the cosmopolitan, technology-based lifestyle, and thus to pipelines and the oil industry. Judge Berger's report, however, explains how there is a very important continuing reliance upon traditional resources. Indeed, without modern equipment, including rifles and snowmobiles, local people would find it 'virtually impossible to continue their traditional land-based subsistence activities'. In some instances, this was because government pressure had caused them to settle in villages far removed from their traditional hunting grounds, so they needed snowmobiles for transport.

'The evidence heard at the Inquiry has led me to conclude that the selective adoption of items of western technology by the Dene and the Inuit [Eskimo] is, in fact, one of the most important means by which they continue to maintain their traditional way of life. These items . . . have become *part of the life* that native people value.' The words which I have italicized here express very precisely two aspects of this kind of interactive innovation. Selective adoption is followed by innovation in the organizational dimension so that new techniques become part of a way of life. This was well illustrated by the organizational changes which the introduction of the snowmobile entailed, but if we use the word interaction to describe this, we should bear in mind how one-sided the process was. Individuals interacted with machines and invented new systems, but few manufacturers have taken note and modified their products to meet local needs.

This is a familiar difficulty in other parts of the world as well. In 1977, a survey of US corporations with branch factories in Africa, Asia, or Latin America commented on how loath they were to put funds and engineering effort into changing product designs to suit local conditions.[9] All the same, some Third World communities do succeed in transforming imported technology to meet their needs, and despite many failures, there has sometimes been effective organizational innovation enabling pumps, tractors and new crops to be incorporated into local cultures.

However, interactive innovation is not always restricted to organization. Sometimes hardware is produced. Sometimes also innovations appear whose significance is chiefly symbolic. Among the latter are the wire toys made all over southern Africa (figure 9). They reflect a local craft tradition, in that African coppersmiths have for centuries produced wire to make bracelets and ornaments – archaeologists have found the draw-plates and other wire-making tools.[10] But these toys adapt wire to the representation of automobiles, often with discarded lids from screw-top jars as wheels, and so represent an innovative response to imported technology. Figure 9 shows a model made by a village boy in eastern Zambia; examples from Malawi, Zimbabwe and Swaziland are based on the same idea. The model is both steered and propelled by means of an extended steering column, and although little attempt is made to represent the vehicle's bodywork, specific details are sometimes reproduced, such as a Mercedes logo and bevel gears made from bottle tops in one Swaziland model. A different kind of wire toy is made for sale to tourists, avoiding the use of bottle tops, without the steering column, and more usually representing bicycles, scooters or even wheelchairs whose tubular framing lends itself to a more realistic form of modeling in wire than the diagrammatic style used for automobiles.

A very different kind of inventiveness was to be seen in Bangladesh during the 1970s, when the bitter struggle for independence seemed to stimulate new awareness of local needs. The result was a stream of ideas for pumps, surveying instruments, transport, and building board

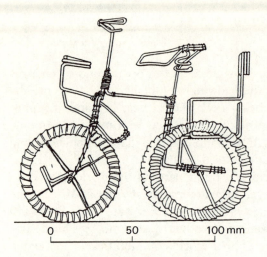

this tricycle was made for sale to tourists by an enterprise known as Zambia Wire Bicyles at Kitwe

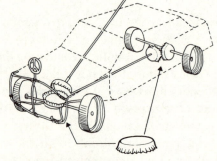

from Swaziland, this toy car has bottle tops to represent bevel gears

made by a village boy from eastern Zambia

FIGURE 9   *Interactive innovation involving the combination of technical ideas from different cultures: craft interest in wire, and elements of industrial technology*
Sources: photographs by Linda Richardson; information from David Farrar, Chris Howes and author's observations. See also *New Civil Engineer*, 19 September 1982, cover illustration which shows an example from Malawi.

from local resources. Two of those responsible, S. S. Ahmed and Najmul Haque, formed a Bangladesh Innovators Association, and Ahmed achieved considerable success with a low-cost duplicating machine using Gestetner stencils. This has all the hallmarks of inter-active innovation. It brought together a wide range of ideas based on local needs (e.g. in schools) plus knowledge of locally available ma-terials (including scrap), and understanding of the imported Gestetner equipment. It was subsequently (1980) manufactured in Bangladesh by Lipikar Industries.

In the world view that predominates today, we tend to ignore oppor-tunities for this kind of interaction, because of linear perspectives that differentiate only between advanced and primitive technology. Innova-tion, we tend to feel, should follow a logical, forward path, and should not be casting back to simpler forms of existing products (as with the duplicator), nor to older traditions, such as survival technology in the Arctic or wire ornament in Africa.

Our linear preoccupations also cause us to ignore the way in which every culture has its own distinctive style in technology-practice, often related to differing organizational procedures and values. Technology is supposed to be universally valid and culturally neutral, and in order to conform with this presupposition, where we do notice differences, we find ways of dismissing them – as most African technology is usually dismissed as of no significance. But there are no easy ways of dismis-sing the distinctive style of technology that may be observed in much of East Asia. Several nations in that region have an approach that is too successful to be ignored, but too different to be understood by a simple linear distinction between levels of sophistication.

The supremacy of East Asia in manufacturing (as opposed to engineer-ing) is no recent development. In 1700, Europeans were attempting to discover how to make porcelain that could match the quality of Chinese and Japanese products. Quality silk and cotton textiles were another challenge difficult to meet. So was paper, and in 1870, a British manufacturer had to devote a considerable research effort to matching the quality of very thin, opaque paper imported from East Asia.[11]

Today, the tradition of quality manufacture and attention to detail is reinforced by other factors. In Japan, the brightest graduates go into production engineering and the consumer goods industries, because there is no NASA and no large defence sector to swallow them up. In other words, there is less scope for the expression of technological virtuosity, but bigger emphasis on economic values. But the Japanese themselves sometimes voice a fear that while they have a good record in innovation, they are less good in original research and invention. Few Japanese have won Nobel prizes. The first transistor radio may have been designed and made in Japan, but the transistor itself originated in the United States. In microelectrics, the Japanese led the way with microprocessor chips of 64K RAM capacity, but feel less confidence about maintaining their lead into the next stage of refinement, the 256K RAM, or with work on the so-called fifth generation computer.

What is especially striking about Japanese industry, of course, is the harmonious working atmosphere; as compared with America, workers 'do not appear angry at superiors and actually seem to hope their company succeeds'.[12] This corporate environment and spirit of consensus makes for efficient production. It makes equally for effective research and development in certain directions. But in a society where, it can seem to westerners, 'nothing is done without arriving at consensus', and individualism is not marked, innovation may more readily follow a linear rather than an interactive pattern. Too ready an acceptance of consensus can inhibit the vigorous discussion on which interaction often depends. Indeed, it is sometimes said that Japanese innovation has been most successful where there have been agreed objectives to work for. It is precisely when development is linear that objectives can be most readily agreed. Thus in Japan, both the value-conflicts and the virtuosity-oriented imperatives which have been so important in the West seem to have less scope.

Whatever may be said for or against this very different style of technology-practice, it does mean that there is immense opportunity for fruitful interaction between Japan and the West. When people talk about microelectronics as the technology of the Pacific rim, they are usually thinking about an economic division of labour, and it is only slowly being appreciated that the mutual stimulus between different styles of technology may be even more significant. The interaction between California's silicon valley, Japan's robotic and computer applications, and the mass production of microprocessors on the Singapore-South Korea seaboard can contribute insight as well as performing an economic function. Even British commentators have begun to talk about the encouraging prospects for 'marrying Japanese manufacturing skills with our strong research and development', mentioning a Japanese link with the ICL computer firm as one strand in an innovative dialogue.

If one difference in the style of technology-practice between Japan

(and China) and the West lies in a greater awareness of the organizational and work aspects, another difference concerns the quality of
production and attention to detail. An American electronic components manufacturer might be happy if the number of defective items in
his output is kept below one per cent of production, whereas a Japanese
corporation will not be content with 0.1 per cent, nor even 0.01 per cent
defective. Similarly, where Japanese-owned factories in Britain have
bought in components from local firms, an unusually high proportion
are rejected. Within these factories also, workers are regularly involved
in meetings for the discussion of quality control.[13] In this and many
other areas, we can recognize, not just problems of competition, but the
value of dialogue and interaction.

A very different example of this sort of dialogue was to be seen in
1974 when a group of American medical men, with US Government
backing, asked themselves whether anything could be learned from
medicine in China that would help them better 'serve the escalating
medical requirements of the American people'.[14] They quickly saw
that the Chinese style of medical practice was too distinctive for much
to be directly transferred to another culture. This was particularly true
of the emphasis on organization and social discipline in the prevention
of disease. Efforts to control malaria, for example, involved holding
numerous meetings in every community to discuss the local problem
and what people could do about it. The whole approach was very
reminiscent of the meetings held in some Japanese factories to alert
workers to problems in quality control.

Thus the discussion about how American medical practice could
benefit from Chinese experience led to a sceptical conclusion about
any direct borrowing of techniques. However, the challenge offered by
the different approaches to be seen in China was readily acknowledged. For example, it was noted that a synthesis between technical
knowledge and an affective, caring approach seemed far more common
than in America, where 'technology gets in the way of caring'. Chinese
medicine was seen to be less virtuosity-oriented and more need-
oriented in another way also: achievements are measured in terms of
the general health of the population, and there is less emphasis on
progress in specialized treatments and techniques.

Dialogue and the 'new professional'

In development projects in the poorer nations of Asia and Africa, the clash of western and local styles in technology-practice often serves to accentuate the divergence between professionals and lay people, experts and users, which was discussed earlier (chapter 3). Very often, the technical experts working on a project will have had a western-style training, and will be separated from the local community not only by professional knowledge and status, but by broader cultural values also. In addition, the disparity in education between professionals and villagers will tempt the former into believing that existing local technology is of little worth, and that their knowledge as experts is a better basis for planning for the future. Interactive innovation is not likely to occur where such attitudes prevail.

An even worse situation arises where technologists, who have no contact with the village, design equipment in distant research institutes, or even in American or British universities, hoping thereby to create appropriate technology. As Michael McGarry says, 'the innovator of the West all too often develops a technology in answer to an imagined problem first, and then proceeds to search overseas for a situation to apply it in'. Such experts necessarily work in ignorance of the people they seek to help; they 'concentrate on hardware, at the expense of social, cultural and organizational aspects. This is a practice proven to be highly susceptible to failure.'[15] As another engineer points out, a successful solar cooker cannot be based only on research concerning solar energy without some study of how people might use it. Yet the user sphere of technology is nearly always neglected or under emphasized in professional research.

The most ironic aspect of the many technological projects that fail because of a lack of any real understanding or dialogue between professionals and people is that the failure is often blamed on the people. They are said to lack willingness to change, and sometimes sociologists are brought in to study the cultural blockages or vested interests that are assumed to be opposed to progress. Yet the real problem is often with the technologist, who has never sat down with people to discover what their lives are about and what they want and need.

But of course, not all experts are like this; some can be identified with the new professionals about whom Robert Chambers writes (chapter 6). Among these, there are people who have worked in a particular locality for so many years that they have inevitably grown into a constructive relationship with the local community, and have evolved an increasing openness to dialogue. Peace Corps volunteers have often been very quick to establish an interactive relationship with local people, seemingly because they are young enough to have a flexible view of professional convention. Sometimes, though, the new professional will be somebody from a research institute who has discovered only by a painful process of trial and error that techniques originating in the laboratory are not always very relevant in the field.

One example concerns grain storage on farms and at village homesteads in Africa. Many different types of traditional granary or silo exist, most of them built with mud walling. However, in some places there has been a heavy loss of grain through the depradations of rats, insects, dampness and mould, and this has contributed to food shortages and malnutrition. Initially it was assumed that such inefficiency was an inevitable part of the traditional technology, which was dismissed as almost worthless. Much effort was therefore devoted to design and trial of concrete or metal silos. In the end, though, granaries built with these materials proved to have few advantages over the traditional ones. They were too costly, and the metal silos tended to over-heat. Meanwhile, the experts had become more aware of the way people used their granaries; where grain was lost, they found, this was due to poor maintenance. What the experts finally did was to accept the merits of the indigenous African designs but suggest detailed improvements that would make maintenance easier.[16]

This is a particularly good example of interactive innovation, with its synthesis of western science, indigenous technology, and expert sensitivity to problems in the user sphere. In the West African state of Mali, the same approach was taken even further. One group of villages in a semi-desert area needed a better water supply. Rainwater could be collected from house roofs, but could not be stored through the long dry season. However, visiting experts were intrigued by the large mud-walled grain bins used in the area, and realized that if ferrocement – that is, cement plastered onto wire reinforcement – were used to strengthen and waterproof them, they would be ideal for water storage.[17] This led to interactive innovation, certainly, and to a dialectical process in an even fuller sense. The men who introduced the ferrocement concept worked out the details with local craftsmen in a collaborative effort. That led them to rethink some of their western values, so that they came to see their work not as modernization but as part of 'the organic development of a traditional society'.

The merit of this approach has been recognized in public health programmes, if nowhere else. The literature on community health discusses at some length how interaction or dialogue may be initiated. The most common approach is to set up village committees in which lay people sit side by side with professionals. The hope is that in such committees, there will be a 'pooling of the expert's knowledge of . . . disease transmission, with the local person's knowledge of local circumstances and behavioural habits'.[18] The success of this depends greatly on the attitudes of the professionals. They may be highly trained for the investigation of narrowly defined problems in medicine or nutrition, but ill-prepared for listening to what may seem just village gossip. They will have a command of highly effective problem-solving skills, but may also be victims of a variety of assumptions and 'blind spots'. Some of the latter may be summarized as follows:

(a) assumptions based on academic specialisms and on boundaries between professions;
(b) the assumption that traditional communities outside the industrialized world have no technology of their own;
(c) a tendency to overlook opportunities for detailed improvements in maintenance and use and to go for technical fixes;
(d) failure to recognize the invisible organizational aspects of technology invariably developed by users of equipment;
(e) failure to recognize the conflicts of values and social goals which specific technological projects may entail.

Robert Chambers suggests two ways in which these obstacles may be overcome by his new professionals. Firstly, existing specialized methods of investigating problems would be combined with a more wide-ranging but non-detailed approach which has variously been described as 'rapid appraisal' and 'taking soundings'. This is a survey method that forces specialists to look beyond their customary disciplinary boundaries; it involves them in collaboration with people from other disciplines as well as with laymen; and it helps to create awareness of the organizational and cultural context of the work.[19]

Secondly, the new professionalism would involve the discipline of reversal mentioned in a previous chapter – that is, deliberately attempting to understand a situation from a point of view opposite to one's normal stance. This sounds vague, but Donald Curtis[20] has suggested an exercise which a professional could undertake to alert himself to what a reversal of values and viewpoints might involve in any particular context; it consists of filling out a matrix that compels one to question accustomed expert views at the same time as lay views and the user sphere are investigated through discussions, surveys and soundings.

However, I would add a third point, for the new professional will not see the sense of either of these approaches unless he has an appropriate view of his role as an expert in society. One suggestion is that the proper role of scientists and technologists is to help draw the maps which society needs in order to steer its future course, but it is not their job to do the steering. This ought to be said of the professional who works with individuals also: when I go to my doctor, I want to know what options are available for dealing with my ailment, but the choice between options I want left in my hands.[21] The difficulty is that all professions involving specialized knowledge are dangerous trades, in that the expert can always present knowledge selectively and manipulate people by pre-empting decisions. He can find himself giving instructions when he should be imparting understanding. Most of us who have professional roles sincerely want to be of service, but we also want recognition and status. Thus, 'many have felt a call to "service" without any idea of becoming "servants". ' We are tempted to use our professional knowledge in exercising overlordship over others; yet the less like servants we are, the less real the service we succeed in giving.[22]

These temptations can present the professional with some of his most deeply-felt dilemmas. An expert on public health may observe that an illness which kills many people in a particular community could be easily prevented if improved latrines were built and used. He may feel such urgency about this that he seeks to impose the technology, 'for the people's own good'. But that is not how a servant of the community should behave, nor in the end is it likely to be of much service. The expert forgets that his sense of urgency arises from a view of life that may have been narrowed by professional training. The people will certainly be concerned about the disease on which he focuses, but they will also be concerned with other problems, related perhaps to unemployment, low income and bad housing. They have to try and cope with all these problems simultaneously, while the expert is concerned with only one of them. And however sensitive and need-directed the expert's primary purpose, he will also have a technical interest in the equipment he proposes, and an inescapable bias towards what seems to him technically sweet. Either way, he is in no position to decide for the people that his solution is the right one for them. All he should do is to put it forward for discussion, advocating his point with all the urgency he feels, but open to counter-suggestions.

Indeed, experience of this kind of situation in countries as diverse as Botswana and Brazil has shown that when latrines are built just on the initiative of professionals, they are not always used for the purpose intended; but where discussion has taken place and the latrines form part of a programme for dealing more broadly with inadequate living standards as the people perceive them – especially housing – individuals will sometimes seize the initiative and build latrines ahead of the programme.

This takes us back to the exercise in reversal which the new professional may undertake to help him see problems from the lay person's point of view. Table 7 represents a greatly modified version of the matrix which Donald Curtis suggested for this purpose. The first question it presents is about the benefits looked for in the project. This encourages the expert firstly to recognize that his own goals refer to very specific benefits, and then secondly to understand the more general but very definite views that local people are likely to have.

Similar questions should be asked about the costs and risks involved, bearing in mind especially the costs which lay people may face in use and maintenance of equipment such as latrines. Adequate cleaning, for example, may absorb time that a busy mother can ill afford, or may entail purchase of cleaning materials. Questions of status will also be important, and not only to the expert who looks forward to the recognition he will gain when the completed project is written up for his professional journal; in some communities, possession of a shiny modern latrine is as much a status symbol for the householder as a new automobile may be in the West.

In many instances where development projects in Africa or Asia fail to make progress, we have already seen that the fault is as likely to be with the expert promoters as with the people. One advantage gained by filling out table 7 is that the promoter is forced to question whether the problems are really due to uncooperative attitudes, or whether the people see the project as irrelevant to their needs. The promoter is also invited to question his own attitudes – that is what reversal is all about. Perhaps 'lack of willingness to change' is something that can be said of his narrowly specialist outlook.

However, Chambers suggests that the most important questions are those that ask: who gains and who loses? These are questions that are affected 'by many decisions which appear technical and neutral'.[23] The green revolution in Asia has taught us that decisions taken in agricultural research affect who benefits. The more prosperous farmers

TABLE 7   Matrix for assessing different points of view on any new
technological development (e.g. a public health project)

The columns representing expert and lay (or user) views are initially blank and
are filled in by promoters of the project as a means of testing its appropriate-
ness in the community concerned. The matrix is here shown partially completed;
in practice, both questions and answers will usually need to be more detailed.

Queries	Expert views	User views
Practical benefits and costs		
What benefits are sought?	Very specific benefits (e.g. control of a particular disease)	Better living standards in general, including health, amenity, housing, jobs
What costs, what risks, and what environmental impacts are perceived?	Cost of implementation; risks as a statistic to be weighed against benefits	Costs in time, cash, amenity, organization, risk, seen in personal and family terms
Who gains which benefits? Who loses?		Lowest income groups cannot afford the cash costs
Status and political advantage		
What is the impact of the project in terms of status and prestige?	Visible progress, good for national prestige Professional advance- ment for the experts concerned	Status associated with possession of new household amenity
Who gains or loses status, power or influence?	Some strengthening of central government authority	Some loss of control over lifestyle; fear of bureaucratic power
Basic values		
What is the cultural context?	Scientific/technical; the expert sphere	Domestic/traditional; the user sphere
What are the dominant values?	Technical interest and virtuosity; economic values	Need or user values, family welfare

'can afford and obtain fertilizers, pesticides, irrigation water, and hybrid seeds. To many smaller and poorer farmers these are out of reach.' An individual scientists' decision to work on biological nitrogen fixation may ultimately help small farmers, but if he works on responses to chemical nitrogen, he must know that he helps only those farmers who can afford to buy fertilizer. However, trends in linear innovation, reinforced by the institutions of the chemical industry, tend continually to give greatest emphasis – and reward the scientist more – for work on the chemical option.

Scope for dialogue

Although the divergence between professional technical interests and the lay person's point of view may be more obvious in Asia or Africa, many of the points made here apply equally in the West. Table 7 represents an exercise in reversal which could be undertaken by the promoters of a nuclear power station in Europe or America as much as by the promoters of latrines in Africa. Indeed, the two columns have been filled out in such general terms that many of the comments could apply equally to a public health or a nuclear project. With both, there is the problem that experts see the goals of programmes in much more specific terms than the public, and with both, basic values differ.

Table 7 is also useful in illustrating how dialogue on such matters is frequently curtailed. When new highways, chemical works or power plant are proposed, many very detailed questions about benefits, costs, and risks are asked in technology assessment exercises, and in environmental impact assessments. These questions cover some of the same ground as the top half of the table, but those who ask them tend to assume that objective answers can be given. There is little recognition of the way in which promoters of projects must usually answer the question differently from the lay public. Yet even to present the lay public as a single entity may be over-simple, for consumers, employees of the project and local residents whose amenity is disturbed will have quite different points of view.

A more significant way in which discussion is curtailed, however, is that questions in the bottom half of the table may never be asked at all. It is assumed that decisions about nuclear energy can be rationally made merely by weighing benefits against risks and costs. Thus many arguments that may be more important to some people are ignored or disguised, and debate on cost-benefit issues becomes distorted by the strong feelings that people have on questions that never surface. Thus concern about the risks of nuclear power becomes shrill and hysterical because those who voice them are worried about a lot of other issues as well. Equally, the claims of the promoters are distorted by a range of hidden motives.

Thus in fuel-rich Britain, the case for nuclear power has been based, since 1980, on elaborately documented claims about the cheapness of nuclear electricity. Yet the differences in cost are marginal, and leaked reports from cabinet committees suggest that the factor which weighed most heavily in favour of the present nuclear programme was a wish to reduce the political leverage which the coal miners' union can exert.[24] That may be a very proper cause for concern, but only if it is debated openly. Subjecting it to a cover-up can only confirm fears that the further development of nuclear electricity is part of a process by which political power is being centralized and consolidated.

If we are to have a democratic control over technological development, we need public inquiry and technology assessment procedures that are able to strip off the various disguises which allow fundamental conflicts of basic values and political interests to hide behind utilitarian arguments about benefits and costs. But many technology assessments and most major British inquiries have served mainly to cloak issues with technical detail that is often barely relevant. Commenting on one such inquiry, a writer in the science journal *Nature* observed that: 'Technical decisions as complex as these have a political content, and that content must be isolated and recognized for what it is.'[25] This had not been done.

It is easy to see why such issues are so rarely fully debated. To engage in a genuinely open dialogue is inevitably to share power over the final decision. If this is a decision about granaries or latrines in an African village, and the only people who have to share this power are a couple of professionals and the villagers, the problem is not insuperable. However, if it is the technocrats who manage a nation's energy supplies and the industrial lobbies and government bodies which support them that must share power with the public, then the stakes are much higher and open dialogue much less likely.

This is important not only in the context of democracy, but also in the interests of innovation. Totalitarian structures, I have argued, restrict innovation to linear paths. Dialogue at many levels, from practical modifications of equipment to formal inquiries, may stimulate innovation in new directions and make it more responsive to people's needs and wants. In western societies, trade unions and consumer groups can play a crucial role, by insisting on high safety standards in equipment and on the performance of welfare functions. In the famous instance of the British firm Lucas Aerospace, trade unionists took on an additional role in putting forward a list of one hundred and fifty socially useful products the firm could be manufacturing instead of working chiefly on armaments.

The organization of work is another important area for dialogue, especially as more computerization is introduced. In one office which had allowed staff some flexibility in fixing their working hours for a number of years, the introduction of word processors meant a return to more rigid work patterns. One employee commented: 'all the . . . machines are switched on and off at the same time, and they want . . . a record of all the work we do, so they can monitor how many orders each person is dealing with'.[26] However, computerization could make more flexible work possible. More people could work from home. There could be modification of the sexual division of labour: job-sharing could allow for family shifts, whereby father leaves work early on one day to meet the children from school and make their tea while mother is at work, but next day the roles are reversed.

Another aspect of technology which cannot be left only to the experts is the allocation of funds and other resources. At the Massachusetts General Hospital in 1980, surgeons were planning ahead for six heart transplant operations in a year. However, it was found that this would absorb such resources that there would be fifty fewer heart operations of more conventional kinds. The transplant programme was then turned down by the hospital trustees, who are mostly lay people. They did not question the doctors' technical judgement, but in this instance were able to press a need-oriented view in place of the virtuosity-oriented imperatives of the experts.[27] This brings out one of the many reasons why dialogue is so important: without it, experts are often carried away by enthusiasm for the technical potential of their work, and lose touch with those aspects of human need they are supposed to serve.

Architects more than most technologists tend to be responsive to arguments such as these, perhaps because the house, more than most other products of technology, must reflect both cultural values and the needs of people. But while architects have sometimes been at the forefront of efforts to devise methods by which people can participate in house design – and in city planning also – there is a tendency among a populist fringe to say that an architect must provide only what lay people want. Where this is taken to its limits, it is arguably as irresponsible as dictating from some technocratic eminence what people must have. Where the professional panders to the public, he gives up on the obligation to maintain a dialogue just as much as when he dictates to them.

Such dialogue is vital at several different levels. It is vital as a dialectic between conflicting sets of values. It is important as a means of balancing narrow specialist views against broader insights. It is highly significant in stimulating innovation through interaction, and leading sometimes to the modification of equipment, or sometimes to new adjustments between organization and technique. In all these ways, if dialogue – or interaction – can be encouraged, future innovation may become more relevant to our problems and needs rather than to experts' ideals of the technically sweet.

9
Cultural Revolution

Democracy and information

Two principal issues have emerged from previous chapters, one intellectual and the other political. The intellectual issue concerns the way value systems inform world views, and how they support beliefs about resources, the arms race, the Third World and technology itself. The political issue concerns the totalitarian nature of many of the institutions which control technology; it is associated with the difficulty encountered at almost every level, of opening any real dialogue between experts and users, technocrats and parliamentarians, planners and people. On the government level, the growth of bureaucracy 'has tended to shunt parliament away from the centre of political life. The executive apparatus functions increasingly without adequate political control.'[1] That has led to a widespread sense of political impotence, and some loss of faith in elected government, and so to the growth of protest movements concerned with the environment, the arms race and nuclear energy.

In both Europe and America, the feeling that totalitarian institutions were taking over was forcibly expressed in the unrest of the late 1960s (especially 1968) and the early 1970s, and in response to this there have been many modest reforms. In several countries, legislators have improved their ability to scrutinize bureaucratic action and technology policy (in Britain, since 1979, through strengthened Parliamentary select committees). There have also been moves to reduce the secrecy that surrounds many decisions; citizens' rights of access to some categories of official information have been recognized in law, first in the Scandinavian countries, then by the American Freedom of Information Act (1967), and later in West Germany (1973) and France (1978). In addition, there have been deliberate efforts to open up

public debate on nuclear energy. In Sweden, from 1973, the government encouraged the formation of study circles to examine the nuclear
issue, and some eight thousand of these local citizens' groups became
active. In Britain, a National Energy Conference was held in 1976 as
part of an effort by the minister responsible, Tony Benn, to widen the
scope of public discussion. In Austria, there was an extended campaign
to inform the public on the nuclear energy issue which ended in 1978
with a referendum that halted the nation's nuclear programme.

A study of these developments, commissioned by the European
Economic Community, has led Jean-Jacques Salomon to put forward
an optimistic vision of technology as a European enterprise, carried
forward in an increasingly co-operative spirit by an informed, partici-
pating public.[2] This is a liberal vision firmly rejecting all determinist
concepts and emphasizing technology as a social process, just as open
to democratic control as any other social process – if people will
appreciate it this way. To secure that appreciation, Salomon advocates
more education in science for everybody, and better training for the
professionals with regard to the social and economic aspects of technology. His hope is for a healing of the divide between the two cultures
based on science and the humanities.

This is a progressive vision, but too much is claimed for the more
open decision-making procedures as they so far exist, and the intellectual issues concerning differing world views and values remain
largely untouched. Salomon recognizes that there are important value-
conflicts that cannot be resolved simply by making more information
available, but argues that when vigorous debate takes place and there is
open participation, perceptions are broadened; expert opinions then no
longer appear exclusively technical, but are seen to involve subjective
judgements and political preferences as well as technical fact. Dorothy
Nelkin has described controversies concerning an airport extension
and a nuclear energy plant in the United States, where open debate in
itself exposed the values built into the experts' technical assessments.[3]
There is thus a case for saying that the development of procedures for
participation is by itself forcing a change in intellectual perspectives;
professionals are forced to abandon the esoteric and sacred land of
scientific facts for the real world, where facts and values are mixed.

This may be partly right, but may also lead to a complacent view of
what participation can achieve. Despite the recognition that facts are
always entwined with value judgements, there is still an assumption
that only one kind of technical information is at issue, and that the way

to promote public participation in technology policy is to ensure the information is widely shared and debated. But any knowledge at all presupposes a world view, and the problem about sharing information is that where world views are in conflict, there will be little agreement about what kinds of knowledge are relevant and valid. Thus one problem at present is that although the framework for debate has opened a little, not much real dialogue takes place – because the different viewpoints that need to interact are not recognized. For example, when technologists submit information to parliamentary committees or public inquiries, they have been known to suggest that their particular proposals are the only rational answer to the problem in hand, and that any other would be 'irrational' or even 'illegitimate'. At the same time, a public inquiry may be conducted on the assumption that the very diverse evidence heard can all be related to a single frame of reference. Often this is done by imposing economic reference points on everything, perhaps by assigning a money value to a botani-cally unique habitat[4] or an ancient church.

At other inquiries, one may observe a public demonstration of the intellectual habit we noted earlier where available information is simply not perceived and is effectively destroyed in order to achieve a coherent view. This seems to have happened at the inquiry into Britain's Windscale nuclear reprocessing project in 1977 where evidence that did not fit a particular concept of 'technical fact'[5] was given little weight. As in Dorothy Nelkin's case-studies, the inquiry evidence certainly exposed the values built into technical arguments. But in the absence of any concept of how to accommodate dissident values, the debate was rather like a discussion between the blind and the deaf – people who perceive different kinds of reality and have no way of discovering how they interconnect.

In this respect, commentators[6] have noticed a sharp contrast in concept between the Windscale inquiry and the roughly simultaneous Canadian inquiry into the proposed Mackenzie Valley pipeline, cited previously in chapter 5. Here, Judge Berger's report pointed out that at least three different sets of values had a bearing on the question – values concerned firstly with the northern frontier, secondly with lifestyle and land, and thirdly with wilderness and environment. Having recognized these values, Berger was then able to suggest an agenda for decision-making, indicating that value-conflicts and claims about land should be resolved before a decision about the pipeline could be taken.

Among the European experiments with widened forms of participation, those which seem to have come nearest to making allowance in their procedures for fundamentally different scales of value may perhaps be found in The Netherlands. Six universities there have opened science schops to provide expert advice – and counter-expert advice – to citizens and community groups worried about environmental issues. At government level, proposals for major physical development projects are discussed by advisory groups representing several mutually opposed points of view, and government ministers must reply to their comments before a parliamentary decision is taken.[7] The concept of counter-information implied by this is perhaps the idea we most badly need, to make clear the point that there is no uniquely correct information. Basic observations and measurements can be factual and neutral, but interpretations, future projections, plans and designs never are – neither is the decision about what to observe and measure. All these are rooted in world views and values, and where the latter differ, the same facts will have different meanings (p. 65). Agreed facts about rising carbon dioxide levels in the atmosphere have different meanings for different specialists and there is no agreement about whether there is a major problem here. Agreed facts about pollution or nuclear accidents are reassuring to some people, alarming to others. Counter-information relates partly to different interpretations of the same data; partly, though, it can compensate for the way experts are trained not to perceive, or to ignore some categories of readily available information (pp. 36 and 152).

Even taking a limited view based on economic efficiency, governments, in their own interest, need to listen to 'multiple voices' and take account of 'multiple public views'.[8] The public interest is only rarely unitary; so the exercise of rationality is 'a more complex and variable process than any conceivable amalgam of "expert" inquiries'.[9]

All this is illustrated in a more restricted but equally clear way by the evaluations of household appliances carried out by some consumer organizations. Such research is based on values and assumptions different from those motivating the manufacturers of the equipment, and sometimes generates information that is new to them.[10] Yet quite a number of people feel that the values of the consumer groups are themselves too narrowly related to efficiency, safety and price, and that not enough consideration is given to the relevance of products for low-income groups, or to their environmental impact. Thus it has been said of the British Consumers' Association, publishers of *Which?* magazine, that they are sometimes just as dogmatic and inflexible as any bureaucracy, 'representing only a narrow range of middle class consumer society'. Among other comments is the suggestion that there could be a *Counter-Which?* magazine to assess products less on the basis of technical efficiency than with regard to the environment, social welfare, the Third World, and so on.[11]

Institutions and education

The task of creating more open and democratic forms of technology-practice cannot be limited to establishing procedures through which public opinion may influence policy and planning. There are issues on which people do not want to participate actively but where it is still very desirable for decisions to accommodate widely different points of view arising from different values and frames of reference.

One subsidiary controversy at the Mackenzie Valley inquiry was connected with the engineering design of pipelines laid in arctic soils, and arose from a phenomenon known as frost heave. Calculations and experiments had been done to check the magnitude of the forces this generated, but the results were in dispute. Even on this specialist problem, then, there were conflicts of information and counter-information. Judge Berger noted that: 'Much of the specialist knowledge and expertise that is relevant to these matters is tied up with the industry and its consultants. This situation is untenable . . . Government cannot rely solely on industry's ability to judge its own case.'[12] On a matter such as this, public participation is barely relevant, but decisions still need to be based on diverse and independent research activity.

Berger thus strongly urged the Canadian government 'to make itself more knowledgeable in matters involving major innovative technology'. Much the same advice could be given to the British government in relation to its nationalized industries. On energy matters, for example, the Atomic Energy Authority has for long been in a privileged position in advising the Department of Energy. There is an especial need for its influence to be countered by a strong, independent energy agency concerned with energy use, consumer interests and conservation. There is also a case for devolving much more responsibility to the regional electricity boards in order to diversify innovation, and to encourage initiatives like those of Midlands Electricity (chapter 8).

More important, though, is the need for more public interest research by independent bodies, especially bodies representative of minority and environmental interests, and of low-income groups. As we have seen, such research is already done in science shops and by some consumer groups. In addition, very valuable studies have been made by Greenpeace on chemical waste dumping in the North Sea, by Friends of the Earth on nuclear power, and by World-Watch on deforestation. But beyond the campaigning style of these latter, often aimed mainly at producing short, polemical publications, there is need for public interest studies with a long-term commitment. The Stockholm International Peace Research Institute (SIPRI) is a good example, and public interest research on arms control is perhaps a more urgent need than anything else. We have already seen how the arms race is sustained by phoney intelligence, and counter-information is needed to off-set this. Another example is the Political Ecology Research Group, which since 1976 has specialized on nuclear energy questions, providing an independent consultancy service which is used by a wide variety of people – local community groups, broadcasting media, the Union of Concerned Scientists (of the United States), and the Lower Saxony State Government (in Germany). [13]

One topic on which public interest research can be particularly important is food, drugs and chemicals. Government regulation of the relevant industries is often fairly tight, but is limited by national boundaries, and is evaded when corporations transfer their activities from one country to another. Thus as tobacco advertising is increasingly restricted within the industrialized countries, sales campaigns in the Third World intensify. Agricultural chemicals which are restricted, 'on safety grounds in the rich countries are freely available in the poor, where the risks are greatest'. [14] A drug which is sold in Africa and Asia as a medicine for children, and vigorously promoted there by an American corporation is known to be 'of no value to children – may actually harm them – and the marketing procedures would be forbidden by law in the US'. [15] The public interest group exposing this scandal – Social Audit – also points to abuses in British sales of milk powder, vitamin pills and hair-care products. The size of the problem is illustrated by one country's attempt to control it. In 1982, Bangladesh announced a ban on 1,500 different drugs, of which 237 were described as harmful and the rest as unnecessary.

One query that has been raised is, how far can one go in encouraging research oriented to many different viewpoints before the result is confusion? Raymond Williams, for example, suggests that in a socialist country committed to central planning, it might be reasonable to suggest that 'there should never be less than two independently prepared plans'.[16] For people used to linear modes of thought, this sounds like a recipe either for disaster or complete paralysis. But if there is to be any democracy in decision-making, alternatives have to be explicit and fully researched. This is a prerequisite for choice. And it need not lead to confusion if alternative views interact in a dialectic of mutual adjustment. If this were recognized, in nations of whatever political complexion, public interest research and critical science would be seen to have a validity and importance in their own right, and would attract support from research councils and foundations. As it is, such work is mainly perceived as a form of opposition to private corporations and government, and thus gains little support from official quarters.

In his vision of a co-operative European commitment to technology based on public participation and a free flow of information, Jean-Jacques Salomon does not fully confront this issue. But he does comment usefully on the importance of better education in science and technology, not only for the citizen but also for the professional. In particular, he makes the point that unless professional technologists are more aware of the socio-economic implications of their work, they will remain locked in illusions of value-free technical rationality, believing that there is only one right answer to every problem. And holding those views, they will not understand how public choice and participation in decision-making can ever make sense.

One may see in detail how the idea of value-free rationality has been perpetuated simply by looking at the textbooks from which many among the present generation of engineers were taught. Most are strongly directed towards the concept of technology as a problem-solving discipline capable of finding 'optimum solutions' and 'right answers' to strictly technical problems. For example, the textbooks from which I was supposed to learn soil mechanics[17] discussed the design of embankments, dams, foundations and highways almost entirely without giving examples of real dams or highways, so questions of context and socio-economic background could never arise. Text-book writers also favoured a very formal style which emphasized the internal logic of the subject. Their model seemed to be Euclid's geometry, with its definitions and axioms, and its logical build-up of theorem upon theorem. An example which follows this pattern very closely, taking Newton's laws of motion as its axioms, is a text on 'mechanical technology' used in teaching technicians for the Ordinary

National Certificate.[18] In this work, the abstraction employed in the effort to seem totally rational and value-free is taken to such extremes that no real machine is mentioned, and engine components such as fly-wheels are referred to only as 'rigid bodies'.

Thought about in this way, technology is quite literally neutral – one might even say sterile. Not surprisingly then, some engineers speak of their formative years as a mind-dulling, disabling experience; it is they who have used the term tunnel vision[19] and in one extreme case, have talked about the need for an engineers liberation' movement.[20] Samuel Florman refers to the stultifying influence of engineering schools in America, where 'the least bit of imagination, social concern or cultural interest is snuffed out under a crushing load of purely technical subjects'.[21]

In many respects, these problems are now better handled. Better textbooks are available. New professional journals discuss technology-practice and its social context, and in Britain, enhanced or enriched engineering degree courses include an extra year of study with emphasis on management and business studies, longer periods of industrial experience, and in some universities, much more design and project work. Elsewhere, there are new courses on science, technology and society (STS).

But this is only a beginning. The additional training in management studies does not automatically mean that engineers have a more rounded, interdisciplinary approach. It may mean that they simply learn fragments of two disciplines without adequately making connections between them. In fact, a central difficulty for the teacher or textbook writer is that if he discusses social context and organization as a sociologist would, he loses touch with real nuts and bolts and practical technology. But if he presents the technical content of a problem in a conventional manner, there is no satisfactory way of bringing in the organizational aspect at all. In other words, bridges are not built between the two cultures simply by tacking extra subjects onto a conventional technical education. The whole philosophy of such training has to be rethought, textbooks and all, in order to present an integrated vision of technology-practice rather than a tunnel vision focused only on its technical aspects.

One way of overcoming these problems may be found through the new design disciplines which have developed within technology itself. They have led to renewed emphasis on design as part of the training of engineers, and several authors make the point that this helps to tie course content to a social background.[22] Apart from that, however, the design disciplines and the soft systems approach which goes with them can have an influence on the overall structure of teaching. An example is the Open University's very wide-ranging foundation course in technology, which has a mainstream component dealing with issues in technology, and tributary components providing basic teaching in mathematics, materials, chemistry, electricity, and so on.

The aim of presenting an integrated vision of technology which some of these approaches illustrate has been at the centre of my own work for the last dozen years. That needs to be mentioned, because it is this work which has led to the somewhat personal view of technology put forward in this book. Apart from some teaching on the course just mentioned, formative experience has included project work with engineering students; exposure to courses on design technology and agricultural engineering; and multidisciplinary editorial work on subjects ranging from public standpost water supplies[23] to human ecology.[24] In all of this, it has seemed particularly important to find ways of breaking through professional boundaries, to develop broader insights in collectively written texts,[25] and to think about extending the same multidisciplinary approach to field surveys and planning.[26]

Three short practical manuals can also be mentioned as tributaries which have fed the mainstream of the present book.[27] Taken together, they could almost form a companion volume, illustrating what the concept of technology-practice may mean in specific applications, and applying the notion of technology as the management of process to specific problems of husbandry and maintenance. The manuals were also written with the defects of conventional textbooks in mind; where the latter are narrowly technical, the manuals attempt a unified view of how the technical and organizational aspects of water supply or nutrition projects are related; and where textbooks are deliberately abstract, the manuals quote identifiable case-studies from Brazil and Botswana, India and Zaire. Through the case-studies rather than by analysis, the users' viewpoint is brought in, forcing one to notice how technology-practice depends on ordinary people, not just experts, and how experts from different disciplines need to collaborate: engineers with social workers, doctors with horticulturists. The contribution of these small booklets is modest indeed, but I do claim one thing for them: that they show how the philosophy of technology presented in this book may have practical application; it is not just a literary formulation.

Frames of reference

Previous paragraphs suggest a random list of recommendations for minor reforms in technology-practice based on a strengthening of public interest research and improvements in technical education. But these suggestions have implications that run counter to the conventional wisdom. They indicate a plurality of approaches rather than one right answer. They present technical fact and engineering design as expressions of world views and values, not of neutral rationality. Thus for any of these reforms to be carried through with conviction, and for it to have the desired effect, there must be a fairly fundamental shift in the frame of reference we use when thinking about technology and the world in which it is applied. This might entail a fairly small adjustment, like the change in conversational subject of Dahrendorf's analogy (chapter 2). But it may prove to be a sufficiently radical change to merit description as a change in levels of awareness, an awakening to new insights, or even a cultural revolution. In the philosophers' jargon, it might be seen as the adoption of a new paradigm – a new pattern for organizing ideas.

Such terms are relevant because, as was suggested earlier (p. 29), the most fundamental choices in technology are not those between solar and nuclear energy, appropriate and high technique. They are choices between attitudes in mind. We may cultivate an exploratory, open view of the world, or we may maintain a fixed, inflexible outlook, tied to the conventional wisdom, in which new options are not recognized.

This need to pose choices about attitudes, including world views and the concept of technology itself, makes it particularly appropriate to talk about cultural revolution. And although I do not employ this term in quite the same sense as socialists do, for this is clearly a non-socialist book, there is still much to be learned from that direction. Socialists talk about revolutionary awareness. We require this especially with respect to awareness of the possibilities for future progress and choice in technology. Moreover, we require it to be an awareness that is shared by everybody because, as Aldo Leopold says, 'all men, by what they think about and wish for, in effect wield all tools'.[28] Or as the socialist writer Raymond Williams[29] has put it, the 'cultural revolution insists . . . that what a society needs, before all . . . is as many as possible conscious individuals'. This, he makes clear, is a first requirement for countering the trend towards technocratic decision-making.

Williams defines the central task of a socialist cultural revolution as, 'the general appropriation of . . . the intellectual forces of knowledge and conscious decision', for the service of the community; he sees it also as a more 'effective response to . . . general human needs, in care and relationships, and in knowledge and development'. One thinks of the improvements in health achieved in Kerala (chapter 4), partly through the openness of local democracy, and partly through a widening of literacy, and a growing awareness among women especially of what they can do to help themselves. In both respects, this is 'the general appropriation of knowledge'. And as Williams says, the 'cultural revolution . . . will be deeply sited among women or it will not, in practice, occur at all'. Many professionals in technology, I have argued, have tended to be diverted away from the service of basic needs and good husbandry through their greater interest in technological virtuosity. The potential contribution of women – and of insights from some non-western communities – lies in the fact that they represent areas of life where technological virtuosity has not yet become dominant.

To speak about cultural revolution is inevitably to recall the China of Mao Zedong during the late 1960s. It is now usual to decry the events of this cultural revolution, but they are still a relevant example. Mao was disturbed by three widening divisions in Chinese society – between town and country, between worker and peasant, and between mental and manual labour. He hoped to bridge these gaps by making education more practical, by sending students and professional people to work in agriculture, by training peasants as barefoot doctors, and by encouraging dialogue between workers and management in factories. My concern is with some very similar divisions in western technological society, and especially the division between the expert sphere and the user sphere. Many aspects of the Chinese approach are pertinent to this, but whereas Mao sought to bridge the gap by rapid change involving brutal compulsion, I share Williams's view that what we are concerned with is a long revolution based on educational development as well as ideological campaigning.

In this context, it is worth remembering that other crucial phases in the development of western technology arose out of experiences of awakening to new possibilities, some of which may justly be called cultural revolutions. I earlier cited the voyages of Columbus and his contemporaries; the scientific revolution of the seventeenth century can also be looked at in this way. Among more directly relevant examples are the new ideas about economics and production that developed in the eighteenth century and contributed to the organizational innovations of the first industrial revolution. Mary Douglas describes this phase as a 'realization which transfixed thoughtful minds in the eighteenth century and onwards . . . that the market is a system with its own immutable laws'.[30] There was a boldness of illumination about the way this idea was grasped and analysed, and for people engaged in trade and industry, a sense of revelation about the new potential open to them. Reading the correspondence of James Watt's associates and other early industrialists,[31] one may still feel their sense of discovery and see how a shift in awareness concerning socioeconomic organization fed ideas into factory development and engineering innovation.

The cultural revolution that we may find occurring today will also involve the exhileration of discovering new insights. In the 1970s, one could sense this among advocates of the use of solar energy and renewable resources. Today there are other new enthusiasms, such as that which, in 1982, gave Britain more microcomputers in schools and homes, and more telextext users per head of population than any other nation. Such things are bound to change perceptions of technology, and to awaken new awareness of the possibilities open to us, but perhaps only in trivial ways.

A more fundamental aspect of current industrial change is the impact on employment. Between 1971 and 1981, Britain lost 1.3 million jobs in ten major industries ranging from automobiles and chemicals to mining and textiles. By 1990, these same industries could lose another 0.5 million workplaces. But new microelectronics industries seem likely to create less than 0.1 million new jobs by 1990,[32] and the only other major prospects for expanded employment are in the arms industries and construction. Whether the consequences of reduced employment is social unrest, or whether instead there is an erosion of the work ethic and a growing feeling that it no longer matters if one has a job, then either way, the change in perceptions of industry and technology must be vast.

But the central preoccupation of any modern cultural revolution must surely be centred on what one university engineer has described as 'the mainspring of technological misdirection'.[33] This is the impulse to go on inventing, developing and producing regardless of society's needs. The result is that we create systems of organized waste in electricity supply, consumer goods and food production, and above all, in the arms race.

But conventional world views disguise much of this and make it seem logical and necessary; they hide the real nature of the technological imperative. Thus the most important part of any cultural revolution – the biggest shift in perceptions and paradigms – could be a reconstruction of world views so that the irrationality of our present pattern of technological progress is no longer hidden. Before pursuing this point, though, we ought first to ask whether a fundamental change in the technological imperative is possible, even if we were more aware of how that imperative is conventionally disguised?

Some authors advocate change that seems too radical to be credible. They portray western man as an 'unbound Prometheus',[34] crazy about science and machines, and pursuing his white whales of technological achievement wildly and obsessively. They see the historical roots of this attitude in the Judaeo-Christian tradition, with its work ethic and its teachings about man's dominion over nature. And they conceive cultural revolution as a turning away from this outlook to something more contemplative and gentle, drawing on eastern types of wisdom, and on Buddhist insights.

One does not have to be a slavish adherent of the conventional wisdom to take alarm at any such suggestion of a wholesale rejection of western thought. It might involve the rejection of liberal values also, and of enlightenment and reason, and even of the basis of modern advances in health and welfare. To advocate anything that could involve this seems as absurd as the more extreme aspects of the nineteenth century's romantic reaction against industry.

The point is taken, but yet there remains the problem of unrelenting drives in technology that make many of us secretly want an arms race, that make us thrill to the risks of advanced nuclear technology, or which draw us into adventure on the frontiers of environmental conquest. Some eastern cultures demonstrate an avoidance of these particular obsessions, and might help us find a new balance in western thought without necessarily abandoning any part of it. The point here is that there is a doubleness in western attitudes and a dialectic between opposed points of view. Beside the half of us that is fascinated by high technology, there is another half already partially in tune with Buddhism. Beside western man, with his virtuosity drives, there is also western woman, seemingly less enthralled by such impulses. Beside Bacon's comments on science as dominion over natures, there is also Bacon's more insistent view that knowledge should be applied in works of compassion, and 'for the benefit and use of life'. Beside the heroic engineers who have built 'cathedrals, railroads and space vehicles to demonstrate the adventuring spirit of man', there have also been engineers who saw their vocation as a social and humanitarian one, like 'John Smeaton, who stressed "civil" engineering as opposed to the military branch (and) William Strutt, who attempted to create a technology of social welfare applicable in hospitals and homes'.[35]

For those inclined to find all the faults of western civilization in its religious tradition, we may note that the same doubleness of vision is to be found there as well. Beside Christ the King, celebrated by daringly engineered cathedrals, motivating crusades and colonial conquests, there is also Jesus the carpenter, healing the sick, concerned for the hungry, and washing his followers' feet. The challenge we ought to recognize in eastern religions, or in the basic-needs economies of Kerala and Sri Lanka, is a challenge to tip the balance in the West's traditional dialectic from conquest and virtuosity towards a point where we can perhaps feel our kinship with Buddhism, and where the work of women and craftsmen, of meeting needs and caring, becomes much more important.

It is worth making these points in terms that refer to religion and may seem rather literary, because our vision and values, even in this calculating, atheistic age, find their power to move us partly through rhetoric and symbolism. It sometimes seems that it is the most hardheaded engineers who talk most freely about their work as cathedral-building. And in the most urgent of our technological dilemmas, the nuclear arms race, women have altered the whole atmosphere of debate by actions that are both heroic and symbolic. A small party of women camped at the gates of a US Air Force base in Britain throughout one of the coldest winters on record to protest against cruise missiles, and sustaining their protest into a second winter, make clear that it is not sufficient just to look at the issue simply in terms of power politics and technology. A similar awareness was generated on a more restricted scale by the Scandinavian women who carried their protest across Russia in the summer of 1982. Identical protests by men would have commanded much less respect in Britain, and would probably not have gained entry to Russia. When women take the lead, it is widely if intuitively recognized that their action represents a distinctive set of values, and not just the immaturity of some overgrown student cause.

World views and waves of progress

The question of nuclear weapons illustrates nearly all the key issues I have tried to tackle in this book. Not only is this a field where technological imperatives and virtuosity drives have an ample scope, but there are also many questions to be asked about the roles played by professional technologists and by totalitarian organizations. Scientists' pressure groups, defence bureaucracies and large-scale industry wield power almost beyond political control. President Eisenhower warned against it; retired defence experts such as Herbert York and Solly Zuckerman have repeatedly raised the alarm; but the only thing that seems ever to move it is sustained, persistent, continuous, vociferous, peacefully disruptive public campaigning. One may regret the use of extra-parliamentary tactics, but faced with a totalitarian military-industrial system that makes its decisions in an extra-parliamentary way, the people have only this resort if they are to exercise their proper sovereignty.

It has been said that during the last decade, most really big initiatives have come from the people: governments have followed where people have led. With regard to environmental concerns, or ending the Vietnam war, or progress in women's rights, 'what was politically opposed or neglected became so strongly supported by ordinary people that governments were led to treat it as good politics'.[36] Similarly, nuclear energy in the United States has 'been made uneconomic by . . . public protest'.[37] Some such claims can even be made with regard to the limited Test Ban Treaty of 1963. George Kistiakowsky records Britain's early resistance to any such agreement.[38] But Britain had a strong Campaign for Nuclear Disarmament which publicized the dangers of fall-out from nuclear tests; not least because of this pressure, the British government eventually played a constructive part in the negotiations.

But to say all this is not to advocate unilateral disarmament nor even a nuclear freeze. It is merely to point out what difficulties face the public in getting its voice heard; and I mention it as a particular instance where the questions raised earlier about the role of dialogue in technological issues ought to be applied. In order to go beyond this and form an opinion about what level of defence is required, we have to consider another point that applies generally to most technology. This is that the world view we use in deciding what kinds of technique to use is a view which must include perspectives on human organization and their international context as well as specific concepts of technology.

In the age of Columbus, a shift in awareness came to many Europeans as a result of the discovery of a new continent, and due to the circumnavigation of Africa and the expansion of trade with Asia (chapter 2). In today's world, we perhaps need the altered frames of reference that could come through rediscovering these same continents. That would mean ceasing to lump them together in the ugly portmanteau concept of the Third World. Then instead of seeing these countries as full of backward people living a 'soup kitchen' existence, we might find that much may be learned from them, especially from the non-industrialized but culturally rich countries.

If such voyages of rediscovery were ever to reach the Soviet Union, they would certainly confirm that this nation presents a special and serious danger, and that it cannot make sense for the West to carry out any sort of extensive, one-sided disarmament. But we may also discover that the Soviet threat has been partly induced by the West's own policies, and that Russia has been encouraged to behave dangerously by being perpetually distrusted, vilified and spoken of openly as the enemy even in the absence of war.

The West has strong vested interests whose prosperity depends on preparation for war, and the problem we may need to recognize is that in some respects, the United States and Russia need each other, and manipulate each other's hostility in order to justify their commitments to virtuosity-oriented technology. Field Marshal Michael Carver points to the type of reconstruction of ideas required if we are to move away from this situation by quoting the former US ambassador in Moscow, George F. Kennan.[39] The behaviour of the Russian leadership, he says, is partly 'a reflection of our own treatment' of them. If we continue to view the Russians as implacable enemies, dedicated to 'nothing other than our destruction – that, in the end, is how we shall assuredly have them'. To view Soviet Russia as eaten up with an absolute malevolence is to allow 'intellectual primitivism and naiveté' to distort our own frame of reference. A first step in revising our views would be to seek a better understanding of Russian civilization, not just in terms of ideology, but by considering its history, traditions and national experience, noting, perhaps, the marked continuity between Tsarist and Soviet ways.

As 1982 ended, the Soviet Union, under a new leader, made proposals for arms control and reduction which seemed serious and far-reaching. Thus there may be new opportunities to undertake voyages of exploration and understanding. If so, we ought to remember that a major obstacle in previous negotiations has been a secret wish in our own culture to perpetuate a technological arms race. This has been vigorously expressed by the lobbying of some scientists, for example.

Thus along with the responsibility to understand Russian civilization, there is also a responsibility to better understand our own. In part, that means removing the disguises we use to hide the real reasons for much of our technology. We are told, for example, that some of the earliest nuclear weapons were made because 'it would have been contrary to the spirit of modern science and technology to refrain voluntarily from the further development of a new field of research, however dangerous'. So when the utilitarian or military purpose of the work was overtaken by events, that did not mean the project's cancellation. Instead, new grounds were found 'for the political and moral justification of its continuance'.[40] It is this business of inventing reasons to justify research and invention that has created many aspects of the world view we now take for granted. It has been my main purpose to get behind these invented reasons and expose the virtuosity concept of technology which they so often conceal.

But in a short, exploratory book, many parts of the argument are inevitably left incomplete. This is particularly regrettable where the more positive, constructive themes are concerned. One of these is the possibility of articulating the values of end use and basic human need more fully, so that they have greater influence in shaping future technology. Here, the most important point stressed is the role played by women, but a point left unexamined is the connection that ought to exist between need-oriented values and environmental concerns.

More fundamental, however, is the suggestion that the very concept of technology itself is open to revision. In chapter 3, I quoted Zuckerman as saying that technologists have made the world more dangerous simply by doing what they conceive to be their job – especially regarding the development of weapons. If this is so, cultural revolution needs to be carried to the point where these experts conceive their jobs differently, and understand technology differently also. Many people have recognized the problem, and some relevant ideas are in circulation, often under the banner of appropriate technology. But much of the discussion has been incoherent, even rhetorical, and to get beyond this stage and begin to explore new styles for western technology, we need to go further in questioning ideas of what technology is about. Is it mainly about making things? Or is it about managing the natural processes of growth and decay in which we are involved? Given that a balance is needed in engineering between construction and maintenance, and in medicine between cure and prevention, where should that balance be struck? How should we use the concept of technology-practice, with its ideas about the interaction of technical and organizational innovations?

It would be wrong to claim that questions of this sort can lead to a concept so comprehensive as to displace entirely the more conventional view of technology as a quest to innovate and venture, to construct and develop. Again we need to think dialectically: this is not a matter of defeating one concept by another, but of tipping the balance away from the virtuosity concept towards the process view.

Freeman Dyson[41] sees the options in technology as 'a choice of two styles, which I call the grey and the green'. If the grey style is typified by physics, plutonium and bureaucracy, the green is represented by biology, horse manure and community. But he adds that we cannot simply replace all the grey high technology by a green approach and more appropriate technology. We cannot suppose that the ideology of 'Green is beautiful' will save us 'from the necessity of making difficult choices'. If human needs are to be met, we require both grey and green; if they are to be met in a civilized, humane way, we require a continuous, active dialogue, not the one right answer offered by either of the opposite points of view. Nuclear energy is not the one right answer required if all human needs are to be supplied, but neither is its total abandonment.

If this sounds like fence-sitting, let it be said that I personally not only lean toward environmental causes, but have a low-energy, near-vegetarian lifestyle which scarcely requires nuclear energy for its support. But those are my own preferences, and it would be wrong to insist that this kind of lifestyle is the only satisfactory outcome of a decision-making process in which a great diversity of people and organizations must participate. One of the best of many reports on the energy question insists that the first priority in this sphere is open, pluralistic debate – otherwise 'projects may come to be decided either by financial overlords on what are believed to be purely economic grounds, or by scientists and engineers on grounds of "technical sweetness" '.[42] The one right answer and the simple formula are always suspect, whether economicaly motivated or technically sweet, whether monetarist, socialist, or antinuclear. Individually, we must live by the light of our own awareness, while valuing a plurality of view in the community. Openness, democracy and diversity are what will save us, not some environmentalist blueprint, nor any technocractic plan. Again Mao Zedong was right in theory if clumsy in practice, for he spoke about walking on two legs, that is, combining different approaches, including both complex techniques and community enterprises.

One possible interpretation of the context of these debates is that we have experienced four waves of industrial revolution during the last two centuries, and that the recession of the early 1980s is the pause which heralds a fifth (chapter 2). During the recession, technical innovation is proceeding apace, and there seems a good chance that ultimately some cluster of institutional and technical developments will fall into place to provide a new pattern for growth. One may even see, again with Freeman Dyson, what techniques could be involved: 'We shall find the distinction between electronic and biological technology becoming increasingly blurred'. Both deal with the fundamentals of information. In both, solar energy can he harnessed particularly effectively. Both allow us to fulfil many of the needs of industrial society using much less energy and other resources than we do now.

The prospects seem good, but a fifth wave of industrial change is not to be beneficiently achieved simply by letting innovation in microelectronics and biology run its course. There are choices to be made about the social and cultural aspects of new and evolving forms of technology-practice, about the institutions which manage technology, and about how the new techniques are applied in the user sphere. In looking at the possible options, attention may turn to the nations of the Pacific rim, where some of the new technology is currently being developed, and where much of its hardware is manufactured. These nations seem to have had remarkable success, but one may feel that theirs has become an excessively materialist culture, over-emphasizing economic values. Confronted with this criticism from a westerner, one Japanese retorted: 'better a materialist culture than a weapons culture'.[43]

Perhaps, though, we can do better than both through a humanitarian stress on need-oriented values, not just as a cosy idealism, nor as a search for 'one right answer', but as a strengthening contribution to a continuing dialectic. For three centuries, people have been turning to Francis Bacon for ideas about the goals and methods of science and technology; and as we have seen, Bacon was motivated by a 'love of

God's creation . . . pity for the sufferings of man, and striving for innocence, humility and charity'.[44] He felt that knowledge and technique should be perfected and governed in love; and that the fruits of knowledge should be used, not for 'profit, or fame or power . . . but for the benefit and use of life'.

Notes

CHAPTER 1   Technology: Practice and culture

1 M. B. Doyle, *An Assessment of the Snowmobile Industry and Sport*, Washington DC: International Snowmobile Industry Association, 1978, pp. 14, 47; on Joseph-Armand Bombardier, see Alexander Ross, *The Risk Takers*, Toronto: Macmillan and the Financial Post, 1978, p. 155.
2 R. A. Buchanan, *Technology and Social Progress*, Oxford: Pergamon Press, 1965, p. 163.
3 P. J. Usher, 'The use of snowmobiles for trapping on Banks Island', *Arctic* (Arctic Institute of North America), **25**, 1972, p. 173.
4 J. K. Galbraith, *The New Industrial State*, 2nd British edition, London: André Deutsch, 1972, chapter 2.
5 John Naughton, 'Introduction: technology and human values', in *Living with Technology: a Foundation Course*, Milton Keynes: The Open University Press, 1979.
6 Charles Heineman, 'Survey of hand-pumps in Vellakovil . . .', unpublished report, January 1975, quoted by Arnold Pacey, *Hand-pump Maintenance*, London: Intermediate Technology Publications, 1977.
7 Leela Damodaran, 'Health hazards of VDUs? – Chairman's introduction', conference at Loughborough University of Technology, 11 December 1980.
8 Quoted by Peter Hartley, 'Educating engineers', *The Ecologist*, **10** (10), December 1980, p. 353.
9 E.g. David Elliott and Ruth Elliott, *The Control of Technology*, London and Winchester: Wykeham, 1976.

CHAPTER 2   Beliefs about progress

1 Derek J. de S. Price, *Little Science, Big Science*, New York: Columbia University Press, 1963, pp. 10, 29.
2 Anthony Wedgwood Benn, 'Introduction', *The Man-Made World: the Book of the Course*, Milton Keynes: The Open University Press, 1971, p. 13.

3 Chauncey Starr, *Current Issues in Energy*, Oxford and New York: Pergamon Press, 1979, pp. 77, 87, 91.

4 Nicholas Rescher, *Scientific Progress*, Oxford: Basil Blackwell, 1978, p. 178.

5 Gerald Leach, 'Energy and food production', *Food Policy*, 1 (1975), especially p. 64.

6 For use of these terms, see David Dickson, *Alternative Technology and the Politics of Technical Change*, London: Fontana/Collins, 1974, pp. 43–4; Ralf Dahrendorf, *The New Liberty*, London: Routledge, and Stanford (California): Stanford University Press, 1975, p. 14.

7 J. R. Jensma, 'The silent revolution in agriculture', *Progress: the Unilever Quarterly*, 53 (4), 1969, pp. 162–5.

8 W. G. Hoskins, 'Harvest fluctuations and English economic history', *Agricultural History Review*, 16, (1968), pp. 15–45; Carlo M. Cipolla (ed.), *The Fontana Economic History of Europe*, vol. 2, London: Collins/Fontana, 1974, Statistical Tables, pp. 612–15. In plotting the graph in Figure 3, all figures have been reduced to a common base by taking a yield of 10 bushels of wheat per acre as equivalent to 690 kg per hectare, or to a yield ratio of around 4 to 6.

9 The Lawes and Gilbert data are quoted and assessed by Susan Fairlie, 'The Corn Laws and British wheat production', *Economic History Review*, ser. 2, 22 (1969), pp. 109–16. Some untypical and very high yields are quoted by M. J. R. Healey and E. L. Jones, 'Wheat yields in England, 1815–1859', *Journal of the Royal Statistical Society*, ser. A, 125 (1962); these latter are not used in Figure 3, except to establish the direction of trends. Recent data are from Ministry of Agriculture sources published as *Agricultural Statistics, United Kingdom*, 1974, 1978 and other years, London: HMSO.

10 On steam engine efficiency, see Starr, *Current Issues*, p. 78; also Carlo M. Cipolla, *The Economic History of World Population*, Harmondsworth: Penguin Books, revised edn 1964, p. 57; Richard G. Wilkinson, *Poverty and Progress*, London: Methuen, 1973, p. 144.

11 D. S. L. Cardwell, *From Watt to Clausius: the Rise of Thermodynamics in the early Industrial Age*, London: William Heinemann, 1974, pp. 158, 179.

12 Anthony Wedgwood Benn, possibly echoing Engels, 'Introduction', *The Man-Made World*, p. 11; compare Dickson *Alternative Technoloy*, p. 46.

13 D. S. Landes, *The Unbound Prometheus: Technological Change and Industrial Development in Western Europe from 1750*, Cambridge: Cambridge University Press, 1969, p. 65.

14 Charles Babbage's preface in Peter Barlow, *A Treatise on the Manufactures and Machinery of Great Britain*, London: 1836, pp. 50–5.

15 Andrew Ure, *The Philosophy of Manufactures*, London: Charles Knight, 1835, p. 15.

16 D. S. Landes, quoted by Stephen A. Marglin, 'What do bosses do?', in *The Division of Labour*, ed. André Gorz, Hassocks (Sussex): Harvester Press, 1977.

17 Arnold Pacey, *The Maze of Ingenuity*, Cambridge (Mass): MIT Press, 1976, pp. 223–6, 272, 277.

18 Ure, *The Philosophy of Manufactures*, p. 23.

19 Ian Crockett, 'An intermediate technology approach to the design of lathes for small workshops', unpublished third year report, Mechanical Engineering Department, University of Manchester Institute of Science and Technology, 1971.

20 Harry Braverman, *Labour and Monopoly Capital*, New York: Monthly Review Press, 1974, pp. 213–20: pp. 169–70, 180, 182 and 223 are also drawn on in this section.

21 Mike Cooley, *Architect or Bee? The Human/Technology Relationship*, Slough: Langley Technical Services, 1980, pp. 2 and 76.

22 Roy Rothwell and Walter Zegveld, *Technical Change and Employment*, London: Frances Pinter, 1979, p. 117.

23 Albert Cherns, 'Automation . . . How it may affect the Quality of Life', *New Scientist*, **78** (8 June 1978), pp. 653–5.

24 Jacques Ellul, *The Technological Society*, trans. John Wilkinson, New York: Vintage Books, 1964, chapter 2, pp. 74, 89.

25 George McRobie, *Small is Possible*, London: Jonathan Cape, 1981, p. 192.

26 Raymond Williams, *Television: Technology and Cultural Form*, London: Fontana/Collins, 1974, pp. 14–21, 128–9.

27 J. S. Weiner, *The Natural History of Man*, London: Thames and Hudson, 1971, pp. 77–8.

28 Solly Zuckerman, *Beyond the Ivory Tower*, London: Weidenfeld & Nicolson, 1970, p. 129.

29 C. Wright Mills, *The Sociological Imagination*, Harmondsworth: Penguin Books, 1970, p. 101.

30 Leslie Sklair, *Organized Knowledge*, London: Hart-Davis MacGibbon, 1973, pp. 237–8.

31 This argument is put by Nathan Rosenberg, *Perspectives on Technology*, Cambridge: Cambridge University Press, 1976, pp. 240–2.

32 Jeremy Rifkin, *Entropy: a New World View*, New York: Viking Press, 1980, pp. 6, 30.

33 Peter Chapman, *Fuel's Paradise: Energy Options for Britain*, Harmondsworth: Penguin Books, 1979 edn., p. 219.

34 Leslie Hannah, *Electricity before Nationalization*, London and Basingstoke: Macmillan, 1979, p. 136. Other data on the efficiency of steam plant used in plotting Figure 5 comes from Thomas Lean, *Historical Statement of the . . . Steam Engines in Cornwall*, London: 1839; D. B. Barton, *The Cornish Beam Engine*, Truro: Bradford Barton, 1969; D. S. L. Cardwell, *From Watt to*

Clausius: the Rise of Thermodynamics in the Early Industrial Age, London: William Heinemann, 1974; H. W. Dickinson, *A Short History of the Steam Engine*, new edn., London: Frank Cass, 1963; A. J. Pacey, 'Some early heat engine concepts', *British Journal for the History of Science*, **7** (1974), pp. 135–45.

35 Christopher Freeman, John Clarke and Luc Soete, *Unemployment and Technical Innovation*, London: Frances Pinter, 1982, pp. 63 etc. Freeman is also quoted at length on this subject by Rothwell and Zegveld, *Technical Change*, pp. 28–34.

36 Simon Kuznets, *Economic Change: Selected Essays*, London: William Heinemann, 1974, pp. 109–118.

37 R. A. Buchanan, *History and Industrial Civilization*, London and Basingstoke: Macmillan, 1979, p. 151.

38 Dahrendorf, *The New Liberty*, pp. 13–14.

39 The worst instance of linear interpretation is in G. R. Talbot and A. J. Pacey, 'Some early kinetic theories of gases', *British Journal for the History of Science*, **3**, (1966–7), pp. 133–49.

CHAPTER 3 The culture of expertise

1 Lewis Thomas, 'Notes of a biology-watcher: the technology of medicine', *New England Journal of Medicine*, **285** (1971), pp. 1366–8. I am indebted to colleagues in the Open University for bringing this to my attention; see their treatment of the same theme in *Living with Technology: Block 6, Health*, Milton Keynes: The Open University Press, 1980.

2 John Cairns, *Cancer: Science and Society*, San Francisco: W. H. Freeman, 1978, pp. 161–2, 168; compare the same author's attitudes four years later: Jonathan Logan and John Cairns, 'The secrets of cancer', *Nature* (London), **300**, (11 November 1982), pp. 104–5.

3 Robert Chambers, 'Introduction', in *Seasonal Dimensions to Rural Poverty*, eds Robert Chambers, Richard Longhurst and Arnold Pacey, London: Frances Pinter, 1981, p. 4.

4 Amory B. Lovins, *Soft Energy Paths*, New York: Ballinger, 1977, and Harmondsworth: Penguin Books, 1977, pp. 140–1.

5 Paul Herrington, 'Demand, a better basis for the water industry?', *The Surveyor*, 27 September 1974, pp. 14–15; on various estimates of leakage, see Paul Herrington, 'The economic facts of water life', in *Water Planning and the Regions*, ed. P. J. Drudy, London: Regional Studies Association, Discussion Paper 9, p. 43; also Fred Pearce, *Watershed*, London: Junction Books, 1982, p. 48, who quotes larger figures for leakage.

6 Sandy Cairncross, Ian Carruthers, Donald Curtis, Richard Feachem, David Bradley, and George Baldwin, *Evaluation for Village Water Supply Planning*, Chichester and New York: John Wiley, 1980, p. 132; for the

hand-pump figures quoted, see 'Water Decade: First Year Review', *World Water*, **4** (1981), special issue, pp. 14–15.

7 M. M. Bakr and Steven L. Kretschmer, 'Schedule of transit bus maintenance', *Transportation Engineering Journal of ASCE*, **103** (TE 1), pp. 173–181; F. E. McJunkin, *Hand Pumps*, The Hague: International Reference Centre for Community Water Supply, 1977, p. 116.

8 Charles Kerr, 'Editorial', *Waterlines*, **1**(2), October 1982, pp. 2–3; on 'invisible technology', see *New Internationalist*, **103**, September 1981, p. 25.

9 Gordon Harrison, *Mosquitoes, Malaria and Man*, London: John Murray, 1978, p. 234.

10 Editorial, 'How to make dreams untrue', *Nature* (London), **289** (19 February 1981), p. 620; P. H. Levin, 'Highway inquiries: a study in governmental responsiveness', *Public Administration*, **57** (1979), pp. 21–49.

11 Solly Zuckerman, 'The deterrent illusion', *The Times* (London), 21 January 1980, p. 10; a longer version of this paper appeared as 'Science advisers and scientific advisers', *Proceedings of the American Philosophical Society*, **124** (1980), pp. 241–55.

12 George B. Kistiakowsky, *A Scientist at the White House*, Cambridge (Mass): Harvard University Press, 1976, pp. 5, 341; preface by C. S. Maier, pp. xxvi, xl.

13 F. J. Dyson, *Disturbing the Universe*, New York: Harper & Row, 1979, pp. 135–6.

14 Ibid., p. 127.

15 Solly Zuckerman, *Nuclear Illusion and Reality*, London: Collins, 1982, p. 123, and 'The deterrent illusion' (note 11 above). See also Richard Owen, 'Russians press US for nuclear test ban treaty', *The Times* (London), 6 August 1982.

16 Laurence McGinty, 'Neutron bombs: a primer', *New Scientist*, 2 September 1982, **95**, pp. 608–13; Michael Carver, *A Policy for Peace*, London: Faber & Faber, 1982, p. 85.

17 Earl Mountbatten, speech delivered in Strasbourg, 11 May 1979; reprinted in *The Times* (London), 28 March 1980 in an advertisement paid for by the World Disarmament Campaign.

18 *The Guardian*, 30 September and 1 October 1982; on the W82 neutron shell, *Sunday Telegraph*, 1 August 1982.

19 Laurence W. Martin, *The Two-edged Sword*, London: Weidenfeld & Nicolson, 1982, p. 41. This author would disagree with the view of pressures from technologists presented here; see his p. 73.

20 Carver, *A Policy for Peace*, p. 109.

21 Dyson, *Disturbing the Universe*, pp. 144–5.

22 Solly Zuckerman, 'The West must halt the nuclear arms race now', *The*

Listener, **104** (16 October 1980), p. 492; and 'Alchemists of the arms race', *New Scientist*, **93** (21 January 1982), pp. 170–2.

23 Thomas H. Maugh, 'Photochemical smog: is it safe to treat air?', *Science*, **193** (1976), pp. 871–3.

24 Richard Peto, 'Distorting the epidemiology of cancer', *Nature* (London), **284** (1980), pp. 297–300 (this is an extended review of *The Politics of Cancer*, by Samuel Epstein, New York, Sierra Book Club, 1978).

25 Jean Robinson, 'Cancer of the cervix: occupational risks of husbands and wives and possible preventive strategies', in *Pre-clinical Neoplasm of the Cervix*, eds J. A. Jordon, F. Sharp, and A. Singer, London: Royal College of Obstetricians and Gynaecologists, 1982, pp. 11–27.

26 *Nature* (London), **284** (1980), review by Richard Peto; also *Nature* (London), **289** (1981), pp. 127–30, 353–7, and 431–2; articles by S. S. Epstein and J. B. Swartz, by John Cairns, and an editorial.

27 Michael Collinson, 'A low cost approach to understanding small farmers', *Agricultural Administration*, **8** (6), November 1981, pp. 433–50.

28 Robert Lilienfeld, *The Rise of Systems Theory: an Ideological Analysis*, New York: Wiley, 1978, pp. 263–4.

29 W. R. Derrick Sewell, 'The role of perception of professionals in environmental decision-making', in *Pollution: the Professionals and the Public*, eds A. Porteous, K. Attenborough, and C. Pollitt, Milton Keynes: The Open University Press, 1977, pp. 146–8, 150, 158–64.

30 John Ryle, *Changing Disciplines*, London: Oxford University Press, 1948, pp. vi, vii, 19–24, 111–15.

31 Carroll Behrhorst, 'The Chimaltenango development project in Guatemala', in *Health by the People*, ed. K. W. Newell, Geneva, World Health Organization, 1975, pp. 30–52.

32 Jean Robinson, private communication, and John Cairns, *Cancer: Science and Society*, pp. 161–2.

33 John Ryle, *Changing Disciplines*, p. 114.

CHAPTER 4 Beliefs about resources

1 Mary Douglas, *Implicit Meanings: Essays in Anthropology*, London: Routledge & Kegan Paul, 1975, pp. 3–7, 236–43.

2 I am indebted in these paragraphs to Philip Payne (private communication) and also to: Andrew Pearse, *Seeds of Plenty, Seeds of Want*, Oxford: Clarendon Press, 1980 (on the green revolution); Thomas T. Poleman, 'A reappraisal of the extent of world hunger', *Food Policy*, **6** (1981), pp. 236–52; Susan George, *How the Other Half Dies*, revised edition, Harmondsworth, Penguin Books, 1977, pp. 30–1 (for a comment on 'starving' Americans and on the statistics).

3 Michael Watts, 'The sociology of seasonal food shortages in Hausaland',

in *Seasonal Dimensions to Rural Poverty*, ed. Robert Chambers, Richard Longhurst, and Arnold Pacey, London, Francis Pinter, 1981, pp. 201–6.

4　Tony Jackson, *Against the Grain: the Dilemma of Project Food Aid*, Oxford: Oxfam, 1982, pp. 9–11, 93.

5　Tim Josling, 'The world food problem', *Food Policy*, 1 (1975–6), pp. 88–9.

6　André van Dan, 'Should we conserve food like energy?' *Food Policy*, 6 (1981), pp. 276–7.

7　In sub-Sahara Africa, between 1971 and 1980, food production increased by an average of 1.8 per cent annually, but population growth was 2.9 per cent. See *The State of Food and Agriculture 1980*, Rome: UN Food and Agriculture Organization, 1981, p. 8, but note that figures of this sort tend to err on the pessimistic side.

8　Advisory Council on Energy Conservation, *Energy for Transport: Long-term Possibilities*, Energy Paper No. 26, London: Department of Energy, 1978, pp. 3–4.

9　The two arguments based on economic considerations are those of Wilfred Beckerman, *In Defence of Economic Growth*, London: Jonathan Cape, 1974, pp. 42, 219; and Julian L. Simon, *The Ultimate Resource*, Princeton (New Jersey): Princeton University Press, and Oxford: Martin Robertson, 1981.

10　Peter Chapman, *Fuel's Paradise: Energy Options for Britain*, Harmondsworth: Penguin Books, 1979 edn., pp. 21, 141n, etc.

11　R. A. Herendeen, T. Kary, and J. Rebitzer, 'Energy analysis of the solar power satellite', *Science*, 205 (1979), pp. 451–4.

12　Chris Whipple, 'The energy impacts of solar heating', *Science*, 208 (1980), pp. 262–6. This article takes a rather pessimistic view of the length of time required for many solar devices to pay back energy used in their construction.

13　Nathan Rosenberg, *Perspectives on Technology*, Cambridge: Cambridge University Press, 1976, p. 223.

14　Nicholas Georgescu-Roegen, *Energy and Economic Myths: Institutional and Analytical Essays*, New York: Pergamon Press, 1976; I am quoting mainly from the essay in this volume also entitled 'Energy and economic myths'.

15　Nicholas Georgescu-Roegen, 'The steady state and ecological salvation', *Bio Science*, 27 (1977), pp. 266–70; see also Georgescu-Roegen's *Demain la décroissance*, translated by Ivo Rens and Jacques Grinevald, Lausanne and Paris: Favre, 1979, p. 117. I am also quoting here from Georgescu-Roegen's 'afterword' in Jeremy Rifkin, *Entropy: a New World View*, New York: Viking Press, 1980.

16　Fred Pearce, 'The menace of acid rain', *New Scientist*, 95 (12 August 1982), pp. 419–22, and on North America, p. 423.

17　Simon, *The Ultimate Resource*, p. 332.

18　Council for Science and Society, *Deciding about Energy Policy*, London: CSS (3/4 St Andrews Hill), 1979, p. 89.

19 Stephen Cotgrove, *Catastrophe or Cornucopia: the Environment, Politics and the Future*, Chichester and New York: John Wiley, 1982, pp. 119–20; many of the same points about time-scales are made by Douglas, *Implicit Meanings*, and Simon, *The Ultimate Resource*, p. 334.

20 J. E. Gordon, *The New Science of Strong Materials*, Harmondsworth: Penguin Books, 1968, revised edition, 1975, p. 107.

21 Anthony Tucker, 'The folly of the chain-saw massacre', *The Guardian*, 6 May 1982; S. L. Sutton, T. C. Whitmore and A. C. Chadwick, *Tropical Rain Forest, Ecology and Resource Management*, Oxford: Blackwell Scientific Publications, 1983, forthcoming; Gerald O. Barney, ed., *The Global 2000 Report to the President: Entering the Twenty-first Century*, Washington DC: US Government Printing Office, 1980, vol. 2, p. 219.

22 John Stansell, 'Renewable energies as bad as nuclear war', *New Scientist*, **91** (20 August 1981), p. 461.

23 I include in this visitors ranging from W. J. Burchell around 1820 to my own visit in 1972; see W. J. Burchell, *Travels in the Interior of Southern Africa*, London, 2 vols., 1822–4.

24 Naomi Mitchison, *Return to the Fairy Hill*, London: William Heinemann, 1966, p. 86.

25 D. R. Gwatkin, 'Food policy, nutrition planning and survival – the cases of Kerala and Sri Lanka', *Food Policy*, **4** (1979), pp. 245–58. The ultimate source for much data used by Gwatkin (and other authors) is *Poverty, Unemployment and Development Policy with special reference to Kerala*, Geneva, United Nations Department of Social and Economic Affairs, 1975.

26 Thomas McKeown, *The Role of Medicine*, Oxford: Basil Blackwell, 1979, pp. 29–77; Arnold Pacey, 'Hygiene and Literacy', *Waterlines*, **1** (1), July 1982, pp. 26–9.

27 Everett M. Rogers, *Modernization among Peasants*, New York: Holt, Rinehart and Winston, 1969, p. 68.

28 Ibid., pp. 56, 82, 91.

29 Catherine Goyder, 'Voluntary and government sanitation programmes', in *Sanitation in Developing Countries*, ed. Arnold Pacey, Chichester and New York: John Wiley, 1978, p. 165.

30 Leela Gulati, *Profiles in Female Poverty*, Delhi: Hindustan Publishing Corporation, 1981, and Oxford: Pergamon Press, 1982, p. 41.

31 Zoë Mars, *Small-scale Industry in Kerala*, Brighton: Institute of Development Studies, Discussion Paper 105, March 1977, pp. 17, 29, 32 and 37.

32 Francine R. Frankel, *India's Green Revolution*, Princeton (New Jersey): Princeton University Press, 1971, pp. 146–8, 155–6. From 1957 and into the 1970s, the communists consistently attracted 30–40 per cent of the poll in Kerala state elections, but this was divided between the Communist and Communist (Marxist) Parties after 1964. Communist Party members

have played a leading part in coalition governments several times since 1957.

33 Marcus Franda, *India's Rural Development*, Bloomington: Indiana University Press, 1979, pp. 156–7.

34 A. Aiyappan, *Social Revolution in a Kerala Village*, London: Asia Publishing House, 1965, p. 95.

35 Gwatkin, 'Food Policy', *Food Policy*, 4.

36 *The State of Food and Agriculture 1980*, per capita food production rose by 2.5% annually during 1971–80.

37 John Fei, Gustav Ranis and Shirley Kuo, *Growth with Equity: the Taiwan Case*, New York and Oxford: Oxford University Press, 1980; these authors claim that the most important improvement in income distribution in Taiwan arose from land reform (p. 38).

CHAPTER 5 Imperatives and creative culture

1 Stephen Cotgrove, *Catastrophe or Cornucopia: the Environment, Politics and the Future*, Chichester and New York: John Wiley, 1982, p. 68.

2 On efficiency as a value, see for example, Peter Chapman, *Fuel's Paradise*, Harmondsworth: Penguin Books, 1975, reprinted 1979, p. 216.

3 Jacques Ellul, *The Technological Society*, trans. John Wilkinson, New York: Vintage Books, 1964, pp. 14, 74.

4 *There's No Future without IT* (booklet), London: Information Technology Year Secretariat, 1982.

5 Dennis Gabor, *Innovations: Scientific, Technological and Social*, New York and Oxford: Oxford University Press, 1970, p. 9.

6 Cotgrove, *Catastrophe*, p. 68.

7 Samuel C. Florman, *The Existential Pleasures of Engineering*, New York: St. Martin's Press, 1976, p. 101; also pp. 60–1.

8 Herman Melville, *Moby Dick*, London: Richard Bentley, and New York: Harper & Row, 1851, chapter 41.

9 Mary Douglas, *Implicit Meanings: Essays in Anthropology*, London: Routledge & Kegan Paul, 1975, p. 246.

10 Robert M. Pirsig, *Zen and the Art of Motorcycle Maintenance*, London: Bodley Head, 1974, chapter 10.

11 J. K. Galbraith, *The New Industrial State*, 2nd British edition, London: André Deutsch, 1972, chapter 15.

12 Herbert F. York, *The Advisors: Oppenheimer, Teller and the Superbomb*, San Francisco: W. H. Freeman, 1976, pp. ix, 81.

13 Maxwell Fry, *Fine Building*, London: Faber & Faber, 1944, pp. 128–9.

14 Cyril Stanley Smith, 'Art, technology and science: notes on their historical interaction', *Technology and Culture*, 11 (1970), pp. 493–549; see also Smith's *A History of Metallography*, Chicago: Chicago University Press, 1960.

15 D. M. Farrar, private communication, 13 September 1982; for histori-
    cal background and information on the sewer renovation methods and
    materials, see J. W. Sellek and D. M. Farrar, 'Urban sewer renewal', paper
    read to the Institution of Municipal Engineers at Washington, County
    Durham, 14 February 1980 (cyclostyled).
16 P. W. Kingsford, *F. W. Lancester: the Life of an Engineer*, London: Edward
    Arnold, 1960, p. 54; also David P. Billington, *Robert Maillart's Bridges: the
    Art of Engineering*, Princeton (New Jersey): Princeton University Press,
    1979, p. 108.
17 Florman, *Existential Pleasures*, p. 60.
18 Gabor, *Innovations*, p. 8.
19 Philip Pacey, *In the Elements Free*, Newcastle: Galloping Dog Press, 1983.
    See part 2, 'Flying Gliders'.
20 Martin Hunt and David Hunn, *Hang Gliding*, Sherborne (Dorset):
    Pelham Books, 1979; also D. A. Reay, *The History of Man-powered Flight*,
    Oxford: Pergamon Press, 1977.
21 Alexander Ross, *The Risk Takers*, Toronto: Macmillan and the Financial
    Post, 1978, pp. 158–9.
22 Jack Burton, *Transport of Delight*, London: SCM Press, 1976, pp. 49–50.
    On the engine names quoted, see Barrie Trinder, *The Industrial Revolution
    in Shropshire*, Chichester: Phillimore, 1973, p. 176.
23 Walt Whitman, 'To a Locomotive in Winter', quoted by Kenneth Hopkins
    (ed.), *The Poetry of Railways*, London: Leslie Frewin, 1966.
24 Erik P. Eckholm, *Losing Ground: Environmental Stress and World Food
    Prospects*, Oxford and New York: Pergamon Press, 1976, p. 146; on the
    arctic 'frontier', see Thomas R. Berger, *Northern Fontier, Northern Home-
    land: the Report of the Mackenzie Valley Pipeline Inquiry*, vol. 1, Ottawa:
    Ministry of Supply and Services, 1977, pp. 112–13.
25 Peter Hartley, 'Educating Engineers', *The Ecologist*, **10** (10), December
    1980, p. 352.
26 Francis Bacon, *The Advancement of Learning and New Atlantis*, ed. Arthur
    Johnston, Oxford: Clarendon Press, 1974. The quotation is from *The New
    Atlantis*, written about 1620 and published posthumously in 1626.
27 Herman Melville, *Redburn*, London: Richard Bentley, and New York:
    Harper & Row, 1849, chapter 32. For other technological imagery used by
    Melville, see *Moby Dick*, chapters 37 (railroads and wheels), 44, 108
    (Prometheus), 60 (steam engine), 54, 85 (the Erie Canal), 103, 107
    (cathedrals), 96, 113, 116, 119 (fires and forges), 36 (electricity).
28 Lewis Mumford, in *The Development of Western Technology*, ed. Thomas
    Parke Hughes, New York and London: Macmillan, 1964, pp. 18–19; also
    John Winter, *Industrial Architecture*, London: Studio Vista, 1970.
29 Gavin Stamp, 'Giles Gilbert Scott: the problem of "modernism" ',
    Architectural Design, **49**, (10–11), October 1979, pp. 80–3.
30 K. E. B. Jay, *Britain's Atomic Factories*, London: HMSO, 1954, p. 53.

31 Alvin M. Weinberg, 'Impact of large-scale science on the United States', *Science*, **134** (1961), p. 161.

32 Robert Jungk, *The Big Machine*, London: André Deutsch, 1969, pp. 1–2.

33 Peter Medawar, *The Hope of Progress*, London: Methuen, 1972, p. 116.

34 Joseph Harriss, *The Eiffel Tower: Symbol of an Age*, London: Paul Elek, 1976, pp. 10–15, 25.

35 Robert Jungk, *Brighter than a Thousand Suns*, trans. James Cleugh, London: Gollancz and Hart-Davis, 1958, chapter 12.

36 Robert Walgate, 'Science in France', *Nature* (London), **296**, (25 March 1982), pp. 299–303; also Peter Pringle and James Spigelman, *The Nuclear Barons*, London: Michael Joseph, 1981.

37 Philip G. Cerny (ed.), *Social Movements and Protest in France*, London: Frances Pinter, 1982, p. 218; also 'Power to the People', *The Guardian*, 7 May 1981.

38 Colin Ward, 'Appendix', in *Fields, Factories and Workshops Tomorrow*, by Peter Kropotkin, abridged version, ed. C. Ward, London: Allen and Unwin, 1974, pp. 159–66.

39 Christopher Freeman, *The Economics of Industrial Innovation*, Harmondsworth: Penguin Books, 1974.

40 P. D. Henderson, 'Two British errors . . . some possible lessons', *Oxford Economic Papers*, **29** (1977), pp. 159–205.

41 David Dickson, *Alternative Technology and the Politics of Technical Change*, London: Fontana/Collins, 1974, pp. 70, 182–4, 195.

42 W. R. Niblet, 'Should State Schools teach Religion?', *The Guardian*, 14 September 1967.

43 Pirsig, *Motorcycle Maintenance*, chapters 1 and 10.

44 Florman, *Existential Pleasures*, p. 125.

45 Leslie Hannah, *Electricity before Nationalization*, London and Basingstoke: Macmillan, 1979, p. 134.

46 C. H. Waddington, *The Scientific Attitude*, Harmondsworth: Penguin Books, 1941, p. 74.

47 Kenneth Barnes, preface to the booklet, *There's No Future without IT*, London: Information Technology Year Secretariat, 1982.

48 M. W. Thring, *The Engineer's Conscience*, London: Northgate, 1980, p. 231.

49 Martin J. Wiener, *English Culture and the Decline of the Industrial Spirit, 1850–1980*, Cambridge: Cambridge University Press, 1981, pp. 141, 150.

50 Hannah, *Electricity before Nationalisation*, p. 152.

51 L. T. C. Rolt, *Isambard Kingdom Brunel*, London: Longman, 1957, seventh impression, 1971, pp. 56, 318, 325.

52 James Bellini, William Pfaff, Laurence Schloesing and Edmund Stillman, *The United Kingdom in 1980: the Hudson Report*, London: Associated Business Programmes, 1974, p. 60.

CHAPTER 6   Women and wider values

1 *The Iliad of Homer*, book XVIII, trans. Alexander Pope, in *The Poems of Alexander Pope*, Twickenham Edition, New Haven: Yale University Press, vol. 7, 1967, ed. Maynard Mack, p. 348. Other Homer quotations come mainly from *The Odyssey*, trans. E. V. Rieu, Harmondsworth: Penguin Books, 1946, books II, VI, VII, VIII and XX.

2 M. I. Finley, *The World of Odysseus*, London: Chatto and Windus, 1964, pp. 77–8.

3 Susan Walker, 'Women in Antiquity', *Slate* (London, 9 Poland St.), 8 (July/August 1978), pp. 14–16, quoting also M. F. Lefkowitz and M. Fant, *Women in Greece and Rome*, Toronto, 1977.

4 David Mitchnik, *The role of Women in Rural Development in Zaire*, Oxford: Oxfam, 1972.

5 Ester Boserup, *Women's Role in Economic Development*, New York: St. Martin's Press, 1970, p. 22.

6 M. R. Haswell, *Economics of Agriculture in a Savannah Village*, London: HMSO, Colonial Research Paper No. 8, 1953.

7 Arnold Pacey, *Gardening for Better Nutrition*, London: Intermediate Technology Publications, 1978, p. 7.

8 This much-quoted finding is interestingly discussed by Liam Hudson, *Frames of Mind*, London: Methuen, 1968, p. 18.

9 Ingrid Palmer, 'Seasonal dimensions of women's roles', in *Seasonal Dimensions to Rural Poverty*, eds Robert Chambers, Richard Longhurst and Arnold Pacey, London: Frances Pinter, 1981, p. 198.

10 J. C. Bryson, 'Women and agriculture in sub-Sahara Africa', in *African Women in the Development Process*, ed. Nici Nelson. London: Frank Cass, 1981, p. 37.

11 M. Z. Rosaldo and Louise Lamphere, ed., *Women, Culture and Society*, Stanford: Stanford University Press, 1974, pp. 221–2. But also see Robert Evenson, 'The new home economics', *Food Policy*, 6 (1981) pp. 180–4.

12 Palmer, 'Seasonal dimensions of women's roles', p. 196.

13 Joan A. Rothschild, 'A feminist perspective on technology and the future', *Women's Studies International Quarterly*, 4 (1981), pp. 65–74.

14 Catharine Hodges and Margot Hewson, ' "Engineer" is not a sexist word', *Engineering Today*, 13 November 1978, p. 22.

15 Emily Tomalin, Alison Taylor, Bob Almond and Terry Thomas, 'Engineering design and appropriate technology', *NATTA Newsletter* (Milton Keynes), No. 18, July/August 1982, p. 28.

16 Leslie Hannah, *Electricity before Nationalization*, London and Basingstoke: Macmillan, 1979, pp. 183–4, 203–5. The idealism and enthusiasm of the early years of the Electrical Association for Women is still echoed in its

journal. One member recalls how she achieved an 'all-electric' house in 1928. 'The effect was apparent immediately . . . a cleaner house, less work and a great joy', see Ethel Cameron Richards, '55 years in the EAW', *Electrical Living Journal*, Winter 1982, p. 6.

17 Ruth Schwartz Cowan, 'The "industrial revolution" in the home', *Technology and Culture*, **17** (1), 1976, pp. 1–23. Also see Christine Bose, 'Technology and changes in the division of labor in the American home', *Women's Studies International Quarterly*, **2** (1979), pp. 295–304.

18 Mike Cooley, *Architect or Bee? The Human/Technology Relationship*, Slough: Langley Technical Services, 1980, pp. 42–4.

19 Stephen Cotgrove, *Catastrophe or Cornucopia: the Environment, Politics and the Future*, Chichester and New York: John Wiley, 1982, pp. 37–9.

20 Unpublished report quoted by Oxfam, *Field Directors' Handbook*, Oxford: Oxfam, 1974 edn, section 34, p. 3.

21 Gary Werskey, *The Visible College*, London: Allen Lane, 1978, pp. 220–1.

22 Robert M. Pirsig, *Zen and the Art of Motorcycle Maintenance*, London, The Bodley Head, 1974, chapter 24.

23 Arnold Pacey, *The Maze of Ingenuity*, Cambridge (Mass): MIT Press, 1976, pp. 140–2, 171–2.

24 Carolyn Merchant, *The Death of Nature: Women, Ecology, and the Scientific Revolution*, San Francisco: Harper & Row, 1980.

25 René Dubos, 'The Predicament of Man', Seventh Science Policy Foundation Lecture, delivered at the Royal Society, London, 1971.

26 Victor Papanek, *Design for the Real World*, London: Thames and Hudson, 1972; Paladin edn, 1974, p. 41.

27 Rachel Maines, 'Fancywork: the archaeology of lives', *Feminist Art Journal*, winter 1974–5, pp. 1–3.

28 Norma Broude, 'Miriam Schapiro and Femmage', *Arts Magazine*, **54** (6), February 1980, pp. 83–7.

29 S. E. Finer, *The Life and Times of Sir Edwin Chadwick*, London: Methuen, 1952, reprinted 1970, pp. 439, 448–9.

30 J. F. LaTrobe Bateman, speech on completion of Glasgow waterworks, 1859, quoted by Peter E. Russell, *J. F. LaTrobe Bateman*, M.Sc. thesis, University of Manchester Institute of Science and Technology, 1980.

31 Quoted by George McRobie, *Small is Possible*, London: Jonathan Cape, 1981, pp. 44–5.

32 Meredith W. Thring, *The Engineer's Conscience*, London: Northgate, 1980, pp. 231–2.

33 Robert Chambers, *Rural Development: Putting the Last First*, Harlow: Longmans, 1983 (forthcoming), chapter 7; the biblical reference is to Matthew, 19:30.

34 Ronald Gray, *Goethe: a Critical Introduction*, Cambridge: Cambridge University Press, 1967, p. 177.

35 Freeman J. Dyson, *Disturbing the Universe*, New York: Harper & Row, 1979, pp. 15–16, 171–2 (on Faust), pp. 6–7 (on toys); the Hudson report on Britain (Associated Business Programmes, 1974) put the mythical Faust next to the real Brunel as symbolic figures of comparable significance in technology.

36 Samuel C. Florman, *The Existential Pleasures of Engineering*, New York: St. Martin's Press, 1976, pp. 145, 147; see also Florman's *Engineering and the Liberal Arts*, New York: McGraw-Hill, 1968, p. 83. Other people who have talked about 'toys' are Victor Papanek and Richard Beeching.

37 Quoted in *African Women*, ed. Nelson, p. 37.

38 Pirsig, *Motorcycle Maintenance*, chapter 25.

39 J. R. Ravetz, '. . . Et Augebitur Scientia', in *Problems of Scientific Revolution* (the Herbert Spencer Lectures, 1973), ed. Rom Harré, Oxford: Clarendon Press, 1975, pp. 48–52.

40 Francis Bacon, *Novum Organum*, London, 1620, book I, aphorism 124 (in book I, aphorism 85, Bacon mentions Faustus as a natural magician with a longing for knowledge).

41 This may, of course, be merely a way of giving legitimate appearance to traditional restrictions on women; but it may genuinely reflect the feeling that it is in the humble roles of life, and in service, that one is nearest to God: Matthew 25:35–40; Luke 1:48, 52–3; John 13:4–5.

42 Oxfam Information Department, 'Sri Lanka 6H: Malaria Control Programme', Oxford: Oxfam, 1982.

43 For Buddhist ideas about *dana*, see Melford E. Spiro, *Buddhism and Society: a Great Tradition and its Burmese Vicissitudes*, London: Allen & Unwin, 1971.

44 Sarvodaya Shramadana Movement, *Ethos and Work Plan*, Moratuwa: Sarvodaya Press, 1976, annexure 18.

45 D. R. Gwatkin, 'Food policy, nutrition planning and survival – the cases of Kerala and Sri Lanka', *Food Policy*, 4 (1979), pp. 245–58.

46 A. T. Ariyaratne, *A Struggle to Awaken*, Moratuwa: Sarvodaya Press, 1982, p. 7.

47 Ibid., pp. 8, 23.

48 Pirsig, *Motorcycle Maintenance*; E. F. Schumacher, 'Buddhist Economics' (1966), in *Small is Beautiful*, London: Blond & Briggs, 1973, especially pp. 53–5; Alain Touraine, *La prophétie anti-nucléaire*, Paris: Seuil, 1980, p. 337.

49 Papanek, *Design for the Real World*, p. 64.

CHAPTER 7   Value-conflicts and institutions

1 James Bellini, William Pfaff, Laurence Schloesing and Edmund Stillman, *The United Kingdom in 1980: the Hudson Report*, London: Associated Business Programmes, 1974, pp. 59–60.

2 L. T. C. Rolt, *Landscape with Machines*, London: Longman, 1971, pp. 154–5, 201, 226; on enjoyment of mobility, pp. 172–3.

3 Leo Marx, *The Machine in the Garden: Technology and the Pastoral Ideal in America*, New York: Oxford University Press, 1964, pp. 26–7; and on Jefferson, pp. 127–35, 139, 141.

4 Stephen Cotgrove, *Catastrophe or Cornucopia: the Environment, Politics and the Future*, Chichester & New York: John Wiley, 1982, pp. 29, 37.

5 J. K. Galbraith, *The New Industrial State*, 2nd British edition, London: André Deutsch, 1972, chapter 15.

6 Sir John Hill, quoted by Robert Walgate, 'The Windscale Report', *Nature* (London), **272** (23 March 1978), p. 301.

7 Liam Hudson, *Contrary Imaginations*, Harmondsworth: Penguin Books, 1967, p. 104.

8 Freeman J. Dyson, *Disturbing the Universe*, New York: Harper & Row, 1979, p. 114.

9 F. I. Ordway and M. R. Sharpe, *The Rocket Team*, London, William Heinemann, 1979, pp. 42, 47, 361.

10 Herbert York, *The Advisors: Oppenheimer, Teller and the Superbomb*, San Francisco: W. H. Freeman, 1976, pp. 106, 121.

11 Dyson, *Disturbing the Universe*, pp. 87, 127.

12 Herman Melville, *Moby Dick*, London: Richard Bentley, and New York: Harper & Row, 1851, chapter 37.

13 C. Reich, quoted by Cotgrove, *Catastrophe*, pp. 80–1.

14 Ian Kennedy, *The Unmasking of Medicine*, London, Allen & Unwin, 1981, p. 54.

15 Solly Zuckerman, 'The West must halt the nuclear arms race now', *The Listener*, **104** (16 October 1980), p. 492.

16 Dixon Thompson, 'Technology with a human face – some Canadian examples,' *Summary Proceedings: Human-Scale Alternatives Conference*, Regina (Saskatchewan): University of Regina, 1977, pp. 62–3.

17 David Elliott and Ruth Elliott, *The Control of Technology*, London and Winchester: Wykeham Science Series, 1976; there is a good discussion of the 'servants of power' view on pp. 92–8.

18 Tony Benn, *The Case for a Constitutional Civil Service*, lecture to the Royal Institute of Public Administration, 28 January 1980, Nottingham, Institute for Workers' Control, Pamphlet No. 69, 1980.

19 J. Langrish, M. Gibbons, W. G. Evans, and F. R. Jevons, *Wealth from Knowledge*, London: Macmillan, 1972, pp. 10, 67.

20 Hudson, *Contrary Imaginations*, p. 69.

21 C. H. Waddington, *The Scientific Attitude*, Harmondsworth: Penguin Books, 1941, pp. 19, 109–111.

22 Hal Dunkelman, *Science Policy*, Milton Keynes: The Open University Press, Course TD 342, units 9/10, p. 29.

23 P. D. Henderson, 'Two British errors . . . some possible lessons', *Oxford Economic Papers*, **29** (1977), pp. 159–205.

24 e.g. Raymond Williams, *Problems in Materialism and Culture*, London: Verso Editions and New Left Books, 1980, p. 255.

25 Hudson, *Contrary Imaginations*, pp. 93–118; Cotgrove, *Catastrophe*, pp. 37, 50–3.

26 My understanding of Jefferson is derived from Leo Marx, *Machine in the Garden*, pp. 127–41.

27 George B. Kistiakowsky, *A Scientist at the White House*, Cambridge (Mass): Harvard University Press, 1976, p. 425.

28 Ralf Dahrendorf, *The New Liberty*, London: Routledge & Kegan Paul, 1975, p. 81.

29 Tony Chafer, 'The Anti Nuclear Movement', in *Social Movements and Protest in France*, ed. Philip G. Cerny, London: Frances Pinter, 1982, pp. 202–20.

30 Alain Touraine, 'Political ecology', *New Society*, **50** (8 November 1979), pp. 307–9.

31 Williams, *Problems*, p. 257; for other comments on the divergence between political ecology or environmentalism and the traditional left, see Cotgrove, *Catastrophe*, pp. 89–92.

CHAPTER 8 Innovative dialogue

1 F. J. Dyson, *Disturbing the Universe*, New York, Harper & Row, 1979, pp. 104–110. On the role of individuals and small companies, see also John Jewkes, David Sawers and Richard Stillerman, *The Sources of Invention*, London: Macmillan, 2nd edn, 1969, especially pp. 205–9.

2 Anthony Tucker, 'When the world put the sun in the shade', *The Guardian*, 4 December 1980.

3 Fred Pearce, 'The menace of acid rain . . . Europe and the CEGB', *New Scientist*, **95**, (12 August 1982), p. 422.

4 P. D. Henderson, 'Two British errors . . . some possible lessons', *Oxford Economic Papers*, **29** (1977), pp. 159–205.

5 SERA, *Non-nuclear Energy Options for the UK*, revised edn, London: Socialist Environment and Resources Association, 1980.

6 Hilary Wainwright and Dave Elliott, *The Lucas Plan*, London: Allison & Busby, 1982; the Design Centre exhibition was entitled, 'Designed in Britain, made Abroad', September/October 1981.

7 On Eskimo bio-engineering, see, J. S. Weiner, *Man's Natural History*, London: Weidenfeld & Nicolson, 1971, pp. 197, 225; on ice alloys, see, Maurice Goldsmith, *Sage: a Life of J. D. Bernal*, London: Hutchinson, 1980, pp. 116, 216; also W. D. Kingery, 'Ice alloys', *Science*, **134** (1961), pp. 164–6; on ITK, see, Michael Howes and Robert Chambers, 'Indigenous

Technical Knowledge', *IDS Bulletin* (Brighton, Institute of Development Studies), **10** (2), January 1979, pp. 5–11.

8 Thomas R. Berger, *Northern Frontier, Northern Homeland: the report of the Mackenzie Valley Pipeline Inquiry*, Ottawa: Ministry of Supply and Services, 1977, vol. 1, pp. 110–13. The three sets of values mentioned are referred to on pp. 1, 29, 94–5, and 112–13.

9 USAID survey quoted by M. G. McGarry, 'Appropriate technology in civil engineering', Convention of the American Society of Civil Engineers, San Francisco, 1977.

10 P. S. Garlake, *Great Zimbabwe*, London: Thames and Hudson, 1973.

11 For bibliography on this, see P. Corsi and P. Weindling (eds), *Information Sources in the History of Science and Medicine*, London: Butterworth, 1983, especially p. 46.

12 Ezra F. Vogel, *Japan as Number One: Lessons for America,* Cambridge (Mass): Harvard University Press, 1979, pp. 22, 131.

13 Barry Fox, 'Japan in Britain', *New Scientist*, **92** (3 December 1981), pp. 655–8; also Mary Goldring, 'People of the Pacific Century', BBC Radio 4 series, 3 November–8 December 1982.

14 Arthur Kleinman, Peter Kunstadter, E. R. Alexander and J. L. Gale, eds, *Medicine in Chinese Cultures*, Washington DC: Department of Health, Education and Welfare, 1975 (publication no. (NIH) 75-653), preface and pp. 479–81, 711, 733–4.

15 McGarry, 'Appropriate Technology' (see note 9).

16 Adrian Moyes, *The Poor Man's Wisdom*, Oxford: Oxfam, 1979, p. 12 (a Tanzanian example); also Extension Aids Branch, *A guide to the safe storage of cereals, oilseeds and pulses*, Lilongwe (Malawi): Ministry of Agriculture, 1973.

17 S. B. Watt, *Ferrocement Water Tanks and their Construction*, London: Intermediate Technology Publications, 1978.

18 Alastair White, 'Health extension in phase two of the slow sand filtration project', The Hague, International Water Reference Centre, 1978, unpublished report, p. 11; see also Alastair White, *Community Participation and Education in . . . Sanitation Programmes*, The Hague: International Water Reference Centre, 1981.

19 Robert Chambers, *Rural Development: Putting the Last First*, Harlow (England): Longmans, 1983 (forthcoming), chapter 7.

20 Donald Curtis distinguishes 'administration values' and 'user values'. See *Sanitation in Developing Countries*, ed. Arnold Pacey, Chichester and New York: John Wiley, 1978, p. 175.

21 An opposite assumption that doctors not only do but should decide for patients is made by Arnold S. Relman, 'The new medical-industiral complex', *New England Journal of Medicine*, **303** (1980), pp. 963–9: 'Unlike consumers shopping for most ordinary commodities, patients do not

decide what medical services they need – doctors usually do that for them. Probably more than 70 per cent of all expenditures for personal health care are the result of decisions of doctors.'

22 Leslie Pacey, 'A sermon', *West China Missionary News*, **39** (11), November 1937, pp. 7–9.

23 Chambers, *Rural Development*, ch. 7.

24 The leaked information dates from 23 October 1979; Malcolm Blackmore, 'An infinitely brighter future?', *SERA News*, Spring 1980, No. 22, p. 5.

25 Robert Walgate, 'Mr. Justice Parker and technical fact', *Nature* (London), **272** (23 March 1978), pp. 300–1.

26 *CIS Report: The New Technology*, London: Counter Information Services, 1979, p. 12.

27 Alexander Leaf, 'The MGH trustees say no to heart transplants', *New England Journal of Medicine*, **302** (8 May 1980), p. 1087.

CHAPTER 9 Cultural revolution

1 Stephen Cotgrove, *Catastrophe or Cornucopia: the Environment, Politics and the Future*, Chichester and New York: John Wiley, 1982, pp. 71, 74.

2 Jean-Jacques Salomon, *Prométhée empêtré*, Paris: Seuil, 1982.

3 Dorothy Nelkin, 'The political impact of technical expertise', *Social Studies of Science*, **5** (1975), pp. 35–54.

4 An instance of this at the Cow Green reservoir in northern England; it seemed that rare flowers and industrial production were being measured on a single scale of values. See Cotgrove, *Catastrophe*, pp. 86–7.

5 Robert Walgate, 'Mr. Justice Parker and technical fact', *Nature* (London), **272** (23 March 1978), pp. 300–1; Martin Stott and Peter Taylor, *The Nuclear Controversy: a Guide to the Issues of the Windscale Inquiry*, London: Town and Country Planning Association with the Political Ecology Research Group, 1980.

6 Guild Nichols, *La technologie contestée*, Paris: OECD, 1979, pp. 73–82, compares the Windscale and Mackenzie Valley inquiries.

7 Salomon, *Prométhée*, chapter 4.

8 P. D. Henderson, 'Two British errors . . . some possible lessons', *Oxford Economic Papers* **29** (1977), pp. 159–205.

9 Raymond Williams, *Problems in Materialism and Culture*, London: Verso Editions and New Left Books, 1980, pp. 264–5.

10 Jeremy Mitchell, 'The consumer movement and technological change', in *The Politics of Technology*, eds Godfrey Boyle, David Elliott and Robin Roy, London: Longman, 1977, p. 210.

11 Mitchell, 'The Consumer Movement', p. 212; also David Farrar and Alice Crampin, private communications, 1982.

12 Thomas R. Berger, *Northern Frontier, Northern Homeland: the report of the Mackenzie Valley Pipeline Inquiry*, Ottawa: Ministry of Supply and Services, 1977, vol. 1, p. 21.

13 *PERG, 1977–79: a report on the Activities of the Political Ecology Research Group*, Oxford: PERG, 1979.

14 David Bull, *A Growing Problem: Pesticides and the Third World Poor*, Oxford: Oxfam, 1982, pp. 38, 93.

15 Charles Medawar and Barbara Freese, *Drug Diplomacy*, London: Social Audit, 1982; also Charles Medawar, *Insult or Injury?* London: Social Audit, 1979.

16 Williams, *Problems*, pp. 264–5.

17 G. N. Smith, *Elements of Soil Mechanics*, London: Crosby Lockwood Staples, 3rd edn., 1974: P. L. Capper and W. F. Cassie, *The Mechanics of Engineering Soils*, London: Spon, 5th edn., 1969; T. W. Lambe and R. V. Whitman, *Soil Mechanics*, New York: John Wiley, 1969.

18 G. D. Redford, J. G. Rimmer, D. Titherington, *Mechanical Technology: a two-year course*, London: Macmillan, 1969.

19 Duncan Mara, 'The influence of conventional practice on design capabilities', in *Sanitation in Developing Countries*, ed. Arnold Pacey, Chichester and New York: John Wiley, 1978, p. 75.

20 S. B. Watt, 'Letters', *New Civil Engineer*, 15 June 1978, p. 53.

21 Samuel C. Florman, *The Existential Pleasures of Engineering*, New York: St. Martin's Press, 1976, p. 92.

22 Gordon L. Glegg, *The Design of Design*, Cambridge: Cambridge University Press, 1969; Victor Papanek, *Design for the Real World*, London: Thames and Hudson, 1972; Paladin 1974, p. 39; Bruce O. Watkins and Roy Meador, *Technology and Human Values*, Ann Arbor (Michigan): Ann Arbor Science, 1977, p. 161.

23 *Public Standpost Water Supplies*, The Hague, International Reference Centre for Community Water Supply, Technical Paper 13, 1979.

24 Robert Chambers, Richard Longhurst and Arnold Pacey, *Seasonal Dimension to Rural Poverty*, London: Frances Pinter, 1981.

25 *Water for the Thousand Millions*, written by the Water Panel of the Intermediate Technology Development Group . . . compiled and edited by Arnold Pacey, Oxford: Pergamon Press, 1977.

26 Arnold Pacey, 'Taking soundings for development and health', *World Health Forum*, **3** (1982), pp. 38–47.

27 Arnold Pacey, *Hand-pump Maintenance*, 1977, revised edn., 1980; *Gardening for Better Nutrition*, 1978; *Rural Sanitation: Planning and Appraisal*, 1980, all from, Intermediate Technology Publications, London.

28 Aldo Leopold, *A Sand County Almanac*, New York: Oxford University Press, 1949, reprinted 1979, pp. 67–8.

29 Williams, *Problems*, pp. 255–72.

30 Mary Douglas, *Implicit Meanings: Essays in Anthropology*, London: Routledge & Kegan Paul, 1975, pp. 232–3.

31 See, for example, R. S. Fitton and A. P. Wadsworth, *The Strutts and the Arkwrights*, Manchester: Manchester University Press, 1958.

32 Philip Beresford, 'The high tech jobs harvest', *Sunday Telegraph*, 12 December 1982, reporting data from Warwick University Institute for Employment Research.

33 M. J. L. Hussey, 'Has the 20th century the technology it deserves?', *Journal of the Royal Society of Arts*, **120** (December 1971), p. 4.

34 One of the best available histories of industrialization in Europe is *The Unbound Prometheus: Technological Change and Industrial Development in Western Europe from 1750*, by D. S. Landes, Cambridge: Cambridge University Press, 1969.

35 Arnold Pacey, 'Engineering, the heroic art', in *Great Engineers and Pioneers in Technology*, vol. 1, eds Roland Turner and Steven L. Goulden, New York: St. Martin's Press, 1981, p. xxi.

36 Richard Jolly, 'The Brandt report', *Journal of the Royal Society of Arts*, **130** (December 1981), pp. 45, 55.

37 Editorial, 'Nothing is even more expensive', *Nature* (London), **299** (23 September 1982), p. 287.

38 On Kistiakowsky, see chapter 3 above; also George B. Kistiakowsky, *A Scientist at the White House*, Cambridge (Mass): Harvard University Press, 1976, p. xxvi.

39 Michael Carver, *A Policy for Peace*, London: Faber & Faber, 1982, pp. 97–8, quoting articles by Kennan in the *New York Review of Books* (21 January 1982) and in *Foreign Affairs* (Spring 1982).

40 Robert Jungk, *Brighter than a Thousand Suns*, trans. James Cleugh, London: Gollancz and Hart Davies, 1958; Harmondsworth: Penguin Books, 1960, p. 159.

41 Freeman J. Dyson, *Disturbing the Universe*, New York: Harper & Row, 1979, pp. 196, 227.

42 Council for Science and Society, *Deciding about Energy Policy*, London: CSS (3/4 St Andrews Hill), 1979, pp. 4–5.

43 Making the same comparison in more measured tones, other authors have said that whereas Japan has integrated its technology policy with its economic planning, the US and UK governments have tended to treat technology policy as 'a special aspect of defence, with heavy prestige overtones'; see Christopher Freeman, John Clarke and Luc Soete, *Unemployment and Technical Innovation*, London: Frances Pinter, 1982, p. 199.

44 J. R. Ravetz, *Scientific Knowledge and its Social Problems*, London: Oxford University Press, 1971, pp. 434–6.

Select Bibliography

Although a good deal has been written on technology policy and the organization of technology, the cultural aspect has rarely been considered as a subject for discussion. To indicate different approaches that are possible, it is worth listing a few books which have had a seminal influence on this one. In two respects they show contrasting ways of dealing with the culture of technology.

Firstly, the subject may be described as it is experienced by engineers (Billington, Florman, Glegg, Rolt); by craftsmen (Sturt); by do-it-yourself practitioners (Pirsig); by the general public (Cotgrove, Leo Marx); and by people from non-European cultural backgrounds (in the Berger report).

Secondly, there is a wide range of disciplinary backgrounds on which a study relating to the culture of technology may be based, including design (Glegg); technology policy (Berger, Kennedy); sociology (Cotgrove); literature (Florman, Leo Marx); history (Smith); biography (Rolt); philosophy (Ravetz).

It should perhaps be noted that a number of books which have influenced this study in important ways are omitted from the following list because they deal with organizational and policy aspects of technology, but not significantly with its cultural aspects. Among these are David Dickson's *Alternative Technology* (London: Fontana/Collins, 1974) as well as David and Ruth Elliott's *The Control of Technology* (London and Winchester: Wykeham, 1976). The following, then, are books which include stress on cultural themes such as creativity, beliefs, values and world views.

Berger, Thomas R., *Northern Frontier, Northern Homeland: the report of the Mackenzie Valley Pipeline Inquiry*, Ottawa: Ministry of Supply and Services Canada, 1977 (see volume I).

Billington, David P., *Robert Maillart's Bridges: the Art of Engineering*, Princeton (N.J.) and Guildford (England): Princeton University Press, 1979.

Cotgrove, Stephen, *Catastrophe or Cornucopia: the Environment, Politics and the Future*, Chichester and New York: John Wiley, 1982.

Florman, Samuel C., *The Existential Pleasures of Engineering*, New York: St. Martin's Press, 1976.

Glegg, Gordon L., *The Design of Design*, Cambridge: Cambridge University Press, 1969.

Kennedy, Ian, *The Unmasking of Medicine*, London: Allen & Unwin, 1981 (see especially Chapter 2).

Marx, Leo, *The Machine in the Garden: Technology and the Pastoral Ideal in America*, New York: Oxford University Press, 1964.

Pirsig, Robert M., *Zen and the Art of Motorcycle Maintenance*, London, Bodley Head: 1974, and New York: Bantam, 1975.

Ravetz, J. R., '. . . Et Augebitur Scientia', in *Problems of Scientific Revolution* (the Herbert Spencer Lectures, 1973), ed. Rom Harré, Oxford: Clarendon Press, 1975.

Rolt, L. T. C., *Isambard Kingdom Brunel*, London: Longman, 1957, seventh impression, 1971.

Rolt, L. T. C., *Landscape with Machines*, London: Longman, 1971.

Smith, Cyril Stanley, *A History of Metallography*, Chicago: Chicago University Press, 1960.

Sturt, George, *The Wheelwright's Shop*, Cambridge: Cambridge University Press, 1923, reprinted 1963.

Williams, Raymond, *Television: Technology and Cultural Form*, London: Fontana/Collins, 1974, and New York: Schocken, 1975.

Index

Italics are used to denote key concepts for which definitions are indicated

advertising, 2, 56, 92–3, 106–7, 165
aesthetics, 82–4, 86, 87, 92, 120
Africa, agriculture in, 58–9, 99, 104,
    108; craft traditions, 145, 147, 151; *see
    also* individual countries
agriculture, 14–17, 53, 56–9, 99–101,
    108, 156; *see also* fertilizers, food, green
    revolution
Ahmed, S. S., 147
aircraft, 85, 86, 93, 94, 131
America, 1, 28, 87, 142; *see also* United
    States
appropriate technology, 102, 111, 112,
    137, 150, 176
architecture, 82, 91, 158–9
Arctic, 1, 2, 87, 142–4, 147, 164
Arkwright, Richard, 18, 19
arms control, 40–5, 127, 165, 174, 175;
    see also Test Ban Treaty
arms race, 40, 45, 124–5, 172; scientists'
    lobbying and, 41, 43, 44, 81, 134;
    industrial interests, 42, 133, 171
artist-engineers, 84, 120
atmospheric pollution, 36, 44, 64, 69,
    139, 141
Austria, 161
automobiles, 44, 84, 86, 120–1, 139,
    142; models of, 145, 146

Bacon, Francis, 50, 51, 87–8, 114–16,
    172, 178
Bangladesh, 53, 145, 165
basic needs, 76, 101, 103, 126, 137
Beckerman, Wilfred, 60, 66
Behrhorst, Carroll, 53
beliefs, 13, 28, 55, 77, 160; about
    progress, 13, 23–4, 28; *see also* values,
    world views

Benn, Anthony Wedgwood, 13, 161
Berger, Thomas, 144, 162, 164
Bernal, J. D., 130
Bible, 87, 98, 113, 114, 115
bicycles, 84, 86
bioeconomics, 66, 68
biological resources, 58, 68, 69
biology, 35, 51, 68
biotechnology, 32, 33, 60, 62, 69, 178
Bombardier, Joseph-Armand, 1
Botswana, 70, 154, 168
Braverman, Harry, 21, 23
Brazil, 58, 69, 87, 154, 168
Britain, agriculture, 14–17; computers,
    148, 158, 171; electricity industry,
    29–31, 36, 106, 139–40; industrial
    revolution, 18–20, 31, 32; reasons for
    economic decline, 20, 91, 94, 96, 131,
    141–2
Brunel, I. K., 84, 94–6, 113, 120, 125–6
Buddhism, 92, 115–19, 172, 173
bureaucracy, 117, 130–1, 137, 141, 160

Cairns, John, 35–6, 53
Campaign for Nuclear Disarmament,
    174
Canada, 1–2, 52, 144, 162, 164
cancer, 35–6, 46–8, 53, 65
caring values, 102, 109, 114, 149
Carlson, Chester, 138
Carter, President, J., 69
Carver, Michael, 42–3, 175
cash crops, 75, 99, 102
cathedrals, 87–92, 93, 95, 111, 118, 120
causation, 43–6, 47
CEGB, 39, 40, 139, 141
Chambers, Robert, 112–13, 152, 154
Chapman, Peter, 61, 66

charity, 114, 115, 118, 179
chemical farming, 17, 30, 32, 156, 165
chemical industry, 44, 58, 64, 94, 156;
    safety and, 47, 90, 135, 165
child-care, 71–3, 104, 106, 109, 113
China, 32, 75, 76, 147, 170; medicine in,
    4, 149
choice, availability of, 27–9, 70, 127, 153,
    169; futures and, 33, 67, 77, 178
Christian teaching, 74, 115, 119, 172,
    173; *see also* Bible
civil engineering, 36–7, 52, 80, 93, 173
Cockerell, Christopher, 138
cogeneration, 36, 139–41
Columbus, Christopher, 28, 29, 87, 170,
    175
combined heat and power, *see*
    cogeneration
computers, 7, 22–3, 106–7, 148, 158,
    171
Concorde airliner, 91, 131
conservation, 58, 68, 120, 126, 144; *see
    also* environmental issues
consumer groups, 106, 158, 163–4, 165
control of technology, 43, 127, 133, 157,
    161
cooking, 104, 109, 110
Cooley, Mike, 22, 106–7
Cotgrove, Stephen, 67–8, 107, 121–2
craftsmen, 85, 120–1, 145–7, 151;
    knowledge of, 83, 143; values of, 83,
    97, 108–9, 110
creativity, 49, 78, 91, 94–5, 107–11
Crockett, Ian, 20, 21
cultural aspect of technology-practice, 10, 40,
    48, 78, 89; definition, 5–6, 49; writers
    on, 81, 82, 200; *see also*
    aesthetics, beliefs, creativity, dialectic,
    ethical disciplines, values
cultural revolution, 75, 136, 160,
    169–73, 176
cultural values, 2–3, 10–11, 82, 142–9
culture of technology, *see* cultural aspect
Curtis, Donald, 152, 154

Dahrendorf, Ralph, 33–4, 134, 169
dame, 36, 37, 111, 112, 166
deforestation, 27, 59, 69, 76, 87, 165
democracy, 26, 75, 127, 132–3, 160–6,
    170, 178
democratic value system, 130, 132–4,
    164

design, 22, 49, 110, 118, 142, 167–8
deskilling, 20, 21, 22
dialectic, 112, 118, 166, 172, 177;
    definition, 123, 132; innovation and,
    141, 151, 159
Dickson, David, 91, 93
disease, prevention of, 49, 68, 149, 153;
    see also cancer, malaria, tuberculosis
doctors, *see* physicians
Douglas, Mary, 55, 65, 81, 171
drugs, 74, 102, 165
Dyson, Freeman, 43, 138, 177, 178

ecology, 27, 62, 67, 68; *see also* political
    ecology
economic growth, 31–3, 59, 67, 75–6, 99
economic values, 89, 100–4, 121, 126–8;
    definition, 101, 102; efficiency and, 13,
    30, 78–9, 105; world views and,
    59–60, 62, 66, 67
education, 72–3, 75, 116, 170; of
    technologists, 11, 20, 123, 164, 166–8
efficiency, 13, 30, 78–9, 105, 163
Eiffel Tower, 89, 90, 118, 120
Einstein, Albert, 81, 92
Eisenhower, President Dwight D., 40,
    45, 133–4, 174
Electrical Association for Women, 106
electricity industries, 31, 36, 71, 74,
    90–1; British, 29–30, 39, 64, 94, 106,
    139–41
electricity power stations, as cathedrals,
    88, 91, 92–3; nuclear, 90–1, 93, 157;
    steam turbines in, 29–30, 37, 139
energy analysis, 61–2, 66, 67
energy resources, 27–8, 33, 55, 58–64,
    67
engineering, aesthetic aspects, 82, 84, 89,
    93–5; compared with medicine, 4, 36,
    112; professional boundaries, 52,
    167–8; professional ethics, 95, 109,
    112, 113; tunnel vision in, 10, 38, 167
engineers, perceptions of, 10, 36–8,
    39–40, 123–4; training of, 20, 36–7,
    54, 105, 166–8; women as, 105, 107
environmental issues, 12, 52, 62–4, 156,
    163; *see also* conservation,
    deforestation, pollution, resources
environmental values, 55, 65, 176
environmentalists, 47, 60, 61, 67–8, 120,
    165
Eskimos, 2–3, 143, 144

ethical disciplines, 112–19, 121–2, 125, 155; definition, 112; role in value systems, 121–2, 132, 152–3; *see also* dialectic, values
experience of technology, 5, 55, 79, 84, 103–8; values as response to, 118, 121–2
experts, *see* professionals
expert sphere, 50, 101, 102, 103, 155, 170; definition, 49, 50

factory system, 18–19, 25, 28, 128
famine, 56–7
farmers, *see* agriculture
Faust, 113, 114, 115, 125
female values, 107–8, 110
feminists, 98, 104, 105, 135
fertilizers, 16–17, 30, 56, 58, 154, 156
firewood, 27, 59, 67, 69, 99
flight, 85; *see also* aircraft
Florman, Samuel, 80, 84, 92, 95, 167
fluidized bed boilers, 64, 139, 140
food, consumption, 36, 53, 56, 71, 75–6, 151; processing, 100, 103, 104; production, 14, 36, 53, 56–9, 73–4, 99; resources, 58–9, 151; supply-fix attitudes to, 36, 57–8, 77; *see also* agriculture, cooking, nutrition
forecasts of demand, 39–40, 126–7
France, 19, 20, 32, 134–5, 160; nuclear energy policies, 90–1, 128, 135
Freeman, Christopher, 31, 32
frontier values, 87, 88, 144
fuels, 27, 59, 67, 68–9, 139, 157
future, projections of, 16, 39–40, 67–8, 104, 126–7

Galbraith, J. K., 5–6, 81, 94, 122, 127
Gambia, 99, 100
Gandhi, M. K., 74, 115
garbage, use as fuel, 139, 140; *see also* waste
gas pipeline, 144, 162, 164
genetics, 35, 67, 69
Georgescu-Roegen, Nicholas, 62–3, 66, 67–8, 70
Germany, 20, 32, 93, 134–5, 142, 160, 165; rocket development in, 124, 128, 129
GNP, 102, 103
Goethe, J. W. von, 113

government administration of technology, 93, 157, 160–3; regulation and, 46, 164, 165
grain yields, 14–16, 58; *see also* rice, wheat
granaries, 151
Greek culture, 97–8, 99, 101
Greenland, 1, 143
green revolution, 56, 57, 74, 133, 154
Gropius, Walter, 88
growth, *see* economic growth
Guatemala, 53, 57

halfway technology, 35–7, 44, 50–1, 52
Haslett, Caroline, 106
health, improvements in, 38, 70–2, 105, 170; risks to, 11, 47–8, 90, 135
health care, 4, 40, 74, 75, 115, 116
health education, 37, 105, 115, 116
heart surgery, 91, 102, 112, 142, 158
Helsinki agreement, 45
Herbert, George, 115
high farming, 16–17
high technology, 113, 120, 128, 137; values of, 79, 93–4, 102, 118, 136
Hippocratic oath, 112, 118
Hodgkin, Dorothy, 108
Homer, 97–8, 99, 100
hospitals, 3, 59, 95, 118, 173
humanism, 52, 78, 89, 119
human resources, 69–70, 77
hydrogen bomb, 81, 124
hygiene, 10, 11, 37, 38, 47, 50, 73

immunization, 35, 38
imperatives, *see* technological imperatives
India, 19, 38, 56, 71, 128, 168; hand-pumps in, 8–11, 38; health and mortality, 38–9, 70–2; *see also* green revolution, Kerala
indigenous technical knowledge (ITK), 143
industrial revolution, 18–20, 23, 31–3, 128
industrialization, waves of, 31–3, 174, 178
industry, 27, 47, 141–2, 147–9, 171; goals of, 94, 127–8; military interest, 42, 45, 133, 158, 171; small-scale, 74, 86, 133, 135, 138, 146–7; *see also* chemical industry, electricity industries, factory system

infant mortality, 70–2, 79, 102, 110
information, 35–6, 55–7; freedom of,
    160–4, 166; resources and, 23, 28, 69,
    178
information technology, 93, 171
innovation, 17, 128, 137–48;
    organizational, 19, 24, 25, 144–5;
    responsibility in, 108–9, 111;
    suppression of, 130–1, 138, 141–2; *see
    also* interactive innovation, invention,
    linear innovation
innovative dialogue, 137–42, 146, 149,
    156–9
innovative movements, 17, 29, 31, 86
institutions, democratic, 132–3, 134,
    164–6; mission-oriented, 128–30;
    totalitarian, 127, 129–30, 132, 133–4,
    135, 164
interactive innovation, 133, 140, 142–7,
    151; definition, 142; response to social
    purpose, 25, 84, 139, 141
intermediate technology, 112; *see also*
    appropriate technology
invention, 13, 138; motivation in, 81,
    84–5, 86, 95, 171
Ionides, Michael, 111, 112
Italy, 19, 28, 142

Jacquard's loom, 19
Japan, 141, 142, 178; manufacturing
    tradition, 147, 149; microelectronics,
    32, 33, 148
Jefferson, Thomas, 121, 123, 132

Kennedy, President John F., 41
Kerala, 70–7, 101, 103, 170, 173
Khrushchev, N. S., 40
Kistiakowsky, George, 40, 174
knowledge, biased perception of, 36, 43,
    55–7, 162; political power and, 23,
    134–5; proper use of, 35–6, 114–15,
    179
Kondratieff, N., 32
Kondratieff waves, *see* waves of
    industrialization
Korea, 76, 149
Kuznets, Simon, 31, 32

labour, division of, 20, 22, 23, 101, 104,
    114; *see also* deskilling, women, work
    organization

labour movements, 74, 75, 77, 108, 135,
    148
Lanchester, F. W., 84
land reform, 74, 76, 77
lathes, 20–2
latrines, 73, 115, 116, 153–4
Leonardo da Vinci, 85
Lesotho, 100
lifestyles, 46, 120, 123; Eskimo, 2, 3, 143,
    144; resources and, 28, 61, 63, 176
Lilienthal, Otto, 85
linear innovation, 138, 139, 148, 156;
    definition, 142
linear view of progress, 14, 16, 34, 55, 138,
    147; definition, 14, 23–4; determinist
    implications, 23–4, 29, 31, 130; values
    and, 78, 126, 130, 136
literacy, 71, 72–4, 116, 170
liveware, 6, 49
lobbying, 44, 45, 176
locomotives, 82, 86–7, 120, 121
low-income groups, 70, 77, 163, 165; *see
    also* poverty

machines, aesthetics of, 82, 120, 125;
    human interaction with, 3, 6, 8, 18,
    20–3, 145
machine tools, 20–2, 115
Mackenzie Valley pipeline, 144, 162, 164
Maillart, Robert, 84
maintenance, 36, 48, 49, 50, 68, 83;
    hand-pumps, 8, 10–11; preventive,
    37–8; skills in 22–3; women and, 104,
    105, 109, 113
malaria, 38–9, 46, 53, 72, 115, 149
Malawi, 145, 146
Mali, 151
malnutrition, 46, 52–3, 56, 76, 108, 151
management, *see* labour, organizational
    aspect of technology, process, work
    organization
manufacturing, 147–9
Mao Zedong, 170, 178
Marx, Karl, 117
Marx, Leo, 121, 123
marxism, 74, 117, 118, 131–2, 135–6
Massachusetts General Hospital, 158
materialist values, 79, 118, 120, 121–2,
    178
mechanics, classical, 167; world views
    based on, 62, 66, 109–10

mechanization, 19, 20, 21, 100, 103
medical practice, 3–4, 52–4
medicine, 35–6, 149, 158, 177;
   agriculture and, 4, 53; engineering
   comparisons, 3–4, 36, 112, 127;
   occupational, 47–8, 52; *see also* drugs,
   health, health care, Hippocratic oath,
   physicians, preventive medicine
Melville, Herman, 81, 88
metal-working, 83, 97, 145
microelectronics, 7, 26–7, 32, 148, 178;
   computers and, 22–3, 148, 171
Midlands Electricity Board, 139–41, 164
military-industrial complex, 133–4, 174;
   see also arms race, nuclear arms,
   weapons
mineral resources, 59, 60–2
moon, landings on, 89, 103, 124
morals, 46, 65; *see also* ethical disciplines,
   values
motorcycles, 84, 86, 109
Mountbatten, Earl, 41, 42
movements in technology, 17, 28–9, 33,
   86
multinational corporations, 134, 165
multiple causes, 20, 43–7, 51, 54, 56

NASA, 128, 134, 138, 147
National Coal Board, 139
NATO, 42, 45
nature, conquest of, 12, 66, 78, 91,
   104–5, 172; laws of, 61, 63, 66, 70,
   166; mastering forces of, 65, 84, 86–7,
   89, 93; values and, 65, 87, 104–5; *see
   also* environmental values, frontier
   values
need-oriented goals, 101, 113, 114, 149
need or user values, 101, 103, 108, 112–19;
   basic needs and, 76, 101, 103, 112–13,
   117; definition, 101–3, 116; other
   values compared, 102, 111, 155, 158
Nelkin, Dorothy, 161, 162
Netherlands, 14, 15, 163
neutron bomb, 41, 42, 43
Nobel Prize, 108, 148
nuclear arms, 40–4, 45, 125, 128; test
   bans, 41, 45, 124, 174; weapons
   development, 81, 88, 89, 174
nuclear energy, 39, 60, 128, 131, 177;
   fuel processing plants, 61, 88, 90, 162;
   opposition to, 134–5, 161, 165; power

stations, 12, 135, 156, 157; symbolism
   of, 88, 89, 90–1
nutrition, 6, 76, 105, 116, 168;
   agriculture and, 53, 56, 100, 108; *see
   also* food, malnutrition

occupational illness, 12, 47–8
Odysseus, 98, 102, 114, 125
oil, 58, 59, 60, 65, 139; *see also* energy
   resources
one-dimensional progress, *see* linear view
   of progress
Open University, 168
Oppenheimer, J. Robert, 81, 125, 130
organizational aspect of technology-practice,
   definition, 4–6, 49; innovation in, 19,
   24, 25, 144–5; professionals and, 43,
   51–3, 149–50; work and, 18–23

Pacific rim, 33, 76, 148, 178
parliaments, 39, 130, 160, 163, 174
participation, 75, 127, 132, 161
particle accelerators, 88, 89
Penelope, 97, 98, 102
pesticides, 16, 58, 165
Peto, Richard, 46, 47
physicians, 4, 36, 52–4, 153, 158
Pirsig, Robert, 34, 92, 117
political control, 11, 79, 90, 117, 157;
   dissent from, 134–5, 160, 174;
   exercise of, 26–7, 74–6, 128, 160–3,
   174; evasion of, 12, 43, 80, 128, 160;
   motivation in technology, 48, 79–80,
   92–3; *see also* control of technology,
   democracy
political ecology, 134–5, 136, 165
politics, openness of, 75, 103, 157, 177;
   power relationships, 118, 128, 129,
   155
pollution, 7, 44, 55, 64; atmospheric, 44,
   64, 139, 141, 163; beliefs about, 55,
   63–5; *see also* waste
population, 57, 58, 68, 71, 75, 76
poverty, 57–9, 122; in Britain, 52, 59,
   108; resources and, 57–9, 69, 70, 77
preventive maintenance, 37–8
preventive medicine, 36, 46–7, 51, 52,
   68, 153
problem-solving, 37–8, 51, 126–7, 150,
   166

process concept of technology, 66, 101–3, 106, 177; definition, 68, 104–5; need values in, 102, 104–5; philosophy of, 68, 117, 177; prevention and conservation in, 37–8, 66, 68, 102, 104

production concept of technology, 66, 102; neglect of consumption and end-use, 36–7, 150, 164; technology as making things, 66, 177; *see also* supply-fix, technical fixes

professional culture, 35, 40, 44, 47, 49, 114, 151

professional ethics, 3, 49, 79, 112, 118; and 'new' professionals, 113, 149, 150–1, 152–3

professionals, 35–6, 80, 108, 109; boundaries between specialisms, 50, 51–3, 101, 168; power of, 26, 128, 129; pressure groups, 41, 43, 80, 174, 176; relations with laymen, 51–2, 150–5, 157–9; tunnel vision of, 10, 38, 48, 51, 54, 167

progress in technology, beliefs about, 13, 23–4, 28; measurement of, 13–18, 70, 72, 102; movements in, 24–34, 86; *see also* linear view

Prometheus, 88, 96, 97, 172

proxy issues, 11–12, 90, 172

public interest research, 45, 133, 164–5, 166, 169

public opinion, 117, 134, 157, 161–3, 174; multiple public views, 163, 169, 177–8

pumps, 8–11, 18, 37, 38, 145

quality control, 147, 149
Quebec, 1
quest, technology as, 92, 125, 128, 132, 177

railroads, 30, 86, 87, 94, 111, 121
rain forest, 58, 69
rationality, 24, 78, 89, 117, 129, 163
Ravetz, J. R., 114, 116
recession, 31, 59, 77, 178
recycling, *see* waste recycling
religion, 88, 92, 111, 117, 118, 173; *see also* Bible, Buddhism, Christian teaching
renewable resources, 58, 60, 141, 171; *see also* biological resources, biotechnology, food, solar energy

research, 35, 43, 81, 128–9, 138; *see also* public interest research
resources, 23, 27–8, 55–68, 77; *see also* biological resources, energy resources, human resources, information, mineral resources, renewable resources
reversal, 112, 152, 154, 156; *see also* dialectic
rice, 73, 74, 75, 99, 100
risk, 11, 47–8, 90, 102, 135, 155, 157
road building, 39–40, 93, 115, 116, 117, 166
Robinson, Jean, 47, 48
rocket engineering, 124, 128, 138
Rolt, L. T. C., 120–1, 122
Rothschild, Joan, 104, 112
Russia, *see* Soviet Union
Ryle, John, 52, 54

Salomon, J-J., 161, 166
sanitation, 72, 73, 111, 154
Sarvodaya Movement, 115–17, 118
satellites, 60, 61, 134, 138
Schumacher, E. F., 112, 117
science, 7, 124–5, 130; goals of, 7, 81, 92, 114–15, 178; large-scale, 80, 81, 88, 89; militarization, 80, 108, 134, 174; social responsibility in, 46, 108, 116
science shops, 163, 165
scientific revolution, 110, 114, 170
scientists, 46, 59, 60, 62, 108, 165
service, ideals of, 113, 115, 119, 153
sewers, 83, 111
silicon chips, 26, 148
Simon, Julian, 60, 65
Singapore, 76, 117, 148
SIPRI, 45, 165
Smith, Adam, 28, 117
snowmobiles, 1–4, 8, 9, 86, 143–5
socialism, 135, 165–6, 169
social medicine, 52
solar energy devices, 60–1, 63, 69, 93–4, 138, 150
South Africa, 53
Soviet Union, 40–2, 93, 130, 173, 175–6
space exploration, 89, 93, 120, 124, 134, 138
Sri Lanka, 46, 70–1, 75–6, 103, 115–17, 173
steam engines, 17–18, 26–9, 86, 88, 121
steam turbines, 29–30, 37, 139
style in technology, 147–9, 150, 176–7

supply fix, 37, 41, 42
Swaziland, 145, 146
Sweden, 1–2, 161
symbolism of technology, 88, 89, 91–4, 113
systems maps, 6, 48, 49, 51
systems theory, 6, 9, 51, 129, 165

Taiwan, 76
Tanzania, 59, 99
technical fixes, 10, 39, 126, 136; definition, 7; pollution and, 44, 52, 62; resources and, 56, 62, 63, 66; *see also* halfway technology
technical sweetness, 43, 81, 83, 91, 177
technocratic values, 124–7, 134
technological determinism, 24–8, 79, 88–9, 91
technological imperatives, 78–82, 88, 89, 110, 125–6; definition, 12, 79; hidden aspects of, 79, 81, 86, 172, 176
technological virtuosity, 81, 87–96, 111, 149; *see also* virtuosity values
technology, branches of, *see* agriculture, engineering, medicine; concepts of, *see* process concept, production concept; culture of, *see* cultural aspect; definition, 5–6, 104, 176–7; forms of, *see* appropriate technology, high technology; practice, *see* technology practice
technology-practice, 6, 29–31, 49, 50–1, 167–8; cultural aspect, 5, 10, 78, 82, 102; definition, 4–6, 122; expert and user spheres in, 48–52, 101, 155; organizational aspect, 4–6, 10–11, 18–23; technical aspect, 5, 7, 13, 36; *see also* cultural aspect, style in technology, *etc.*
Teller, Edward, 41, 124–5
Test Ban Treaty, 41, 45, 124, 174
textbooks, 166–7
Thatcher, Margaret, 93
thermodynamics, 62–3, 66, 67, 105
Third World, 37, 111, 133, 145; carica-tures of, 57, 58, 175; drug exports to, 47, 165; food supplies, 36, 56–7, 175; *see also* Africa and individual countries
Thomas, Lewis, 35, 37
totalitarian institutions, 127–31, 133, 135, 157, 160
toys, 113, 137, 145, 146

tractors, 17, 84, 91, 100
trades unions, 75, 106, 135, 157; arms manufacture and, 45, 46, 158; safety issues, 48, 139, 158
transformations of technology, 142, 145
trees, 58, 64, 69, 87
tuberculosis, 52, 53
tunnel vision, 10, 38, 48, 59, 100, 167
Turner, J. M. W., 86

unemployment, 23, 52, 74, 108, 171
United Kingdom, *see* Britain
United States, 1, 32, 53, 86, 106, 148–9; arms race and, 40–1, 81, 128, 175; energy resources, 27, 64, 128, 130, 144; engineering profession, 80, 105, 112, 167; freedom of information, 160, 165; frontier values, 87, 88, 121; grain production, 14–15, 56–7; industrial corporations, 86, 94, 141–2, 145, 165; large-scale science, 88, 124–5, 134, 138; medicine, 149, 158
uranium, 61–2, 90
user sphere, 48–52, 101, 150–1, 155, definition, 49
user values, 101, 102–3, 111, 112, 136; *see also* need or user values

values in technology, 5, 7–12, 78, 101–7; classification of, 101, 102; conflicts of, 89, 120–4, 144, 148; master values, 122, 124–6, 132; reflecting experience, 107, 118, 121–2; responsibility and, 46, 65, 95, 108–12; value-free fallacy, 2, 78, 82, 147, 166–7; world views and, 65, 78, 160, 172–3, 174; *see also* economic values, frontier values, need or user values, virtuosity values, *etc.*
value systems, 120, 122, 124–36; dialectic in 123, 132–3; master values in, 124–6
Vietnam war, 174
virtuosity values, 87–96, 111, 112, 120, 149; adventuring impulse, 84–5, 89, 98, 125–6; definition, 81, 85, 102; economics and, 89, 101–3, 137, 141; male roles and, 98, 114; *see also* frontier values, high technology
visual display units (VDUs), 11–12, 22, 55
Von Braun, Wernher, 124, 125–6, 134, 138

Waddington, C. H., 108, 129–30
waste, 63–4, 135, 171; recycling of, 63, 66, 74, 139, 140; *see also* garbage, pollution
water supplies, 4, 39, 52, 72, 112, 168; neglect of maintenance in, 36–8, 168; *see also* dams, pumps, sanitation
Watt, James, 18, 25, 27, 29, 30, 171
waves of industrialization, 31–3, 174, 178
weapons, culture of, 81, 88, 97–8, 124, 178; conventional, 42, 45; industrial interest in, 42, 45, 133, 147, 171; scientists and 40–1, 43–4, 81, 89, 128; *see also* hydrogen bomb, military-industrial complex, nuclear arms
welfare, 70, 74, 75, 76, 100, 110; *see also* basic needs, health, need or user values, poverty
Weinberg, Alvin, 88, 95
wheat, 14–16
Whitman, Walt, 87
wilderness values, 144, 162
Williams, Raymond, 25, 165, 169, 170
Wilson, Harold, 93
women, 72–3, 97–111; agricultural work of, 17, 73, 99, 100, 104; cultural

revolution and, 170, 173; electricity industry and, 30, 106; experience of technology, 103, 105; health and, 48, 72–3; industrial work, 20, 73, 74; skills of, 97–8, 104–5; values of, 102, 107, 110–11, 115
work camps, 115, 117
work ethic, 171
work organization, 12, 18–23, 50, 158; *see also* labour, factory system
world views, 58, 65–7, 169, 174–7; economics and, 59–60, 62; scientific, 62, 65, 66, 162, 163; values supporting, 65, 78, 160, 172, 174
World War II, 124, 128, 130, 133, 138
Wright, Orville and Wilbur, 85, 86

xerography, 84, 138

York, Herbert, 81, 124, 174

Zaire, 99
Zambia, 145, 146
Zimbabwe, 145
Zuckerman, Solly, 41, 43, 45, 127, 174, 175